AF348399

Advances in Anatomy Embryology and Cell Biology

Vol. 146

Editors

F. Beck, Melbourne   D. Brown, Charlestown
B. Christ, Freiburg   W. Kriz, Heidelberg
E. Marani, Leiden   R. Putz, München
Y. Sano, Kyoto   T. H. Schiebler, Würzburg
K. Zilles, Düsseldorf

Springer-Verlag Berlin Heidelberg GmbH

H.-W. Korf   C. Schomerus   J. H. Stehle

The Pineal Organ, Its Hormone Melatonin, and the Photoneuroendocrine System

With 31 Figures

Springer

H.-W. KORF
C. SCHOMERUS
J.H. STEHLE
Dr. Senckenbergische Anatomie
Anatomisches Institut II
Johann Wolfgang Goethe-Universität
Theodor-Stern-Kai 7
60590 Frankfurt

ISBN 978-3-540-64135-3   ISBN 978-3-642-58932-4 (eBook)
DOI 10.1007/978-3-642-58932-4
Library of Congress-Cataloging-in-Publication Data
Korf, H.-W. (Horst-Werner), 1952– . The pineal organ, its hormone mela-
tonin, and the photoneuroendocrine system / H.-W. Korf, C. Schomerus, J.H.
Stehle. p. cm. – (Advances in anatomy, embryology, and cell biology: V. 146)
Includes bibliographical references and index.

1. Pineal gland-Physiology. 2. Melatonin-Physiological effect. I. Schomerus,
C. (Christof), 1963– . II: Stehle, J.H. (Jörg H.), 1954– . III. Title. IV. Series.
QL801.E76 Vol. 146 [QP188.P58] 573.8'374–dc21

This work is subject to copyright. All rights are reserved, whether the whole
or part of the material is concerned, specifically the rights of translation,
reprinting, reuse of illustrations, recitation, broadcasting, reproduction on
microfilms or in any other way, and storage in data banks. Duplication of
this publication or parts thereof is permitted only under the provisions of the
German Copyright Law of September 9, 1965, in its current version, and
permission for use must always be obtained from Springer-Verlag. Viola-
tions are liable for prosecution under the German Copyright Law.

© Springer-Verlag Berlin Heidelberg 1998
Originally published by Springer-Verlag Berlin Heidelberg New York in 1998

The use of general descriptive names, registered names, trademarks, etc. in
this publication does not imply, even in the absence of a specific statement,
that such names are exempt from the relevant protective laws and regulations
and therefore free for general use.

Product liability: The publishers cannot guarantee the accuracy of any information about dosage and application contained in this book. In every individual case the user must check such information by consulting the relevant
literature.

Production: PRO-EDIT GmbH, D-69126 Heidelberg
SPIN: 10660179   27/3136-5 4 3 2 1 0   – Printed on acid-free paper

Preface

The present contribution on the pineal complex, its hormone melatonin, and the photoneuroendocrine system combines findings from comparative anatomy and physiology with results from cellular and molecular biology. The selection of the topics presented is necessarily subjective and far from complete. Nevertheless we hope to point out some major research developments in recent decades which confirm the functional significance of the pineal organ as an important component of the vertebrate photoneuroendocrine system and its usefulness as a model to study signal transduction mechanisms in neuroendocrine cells and neurons.

Several colleagues have helped us with the preparation of the manuscript. Our special thanks go to Prof. Andreas Oksche (Giessen), who not only provided us with previously unpublished photographs but has guided our research as stimulating mentor for several years. We are grateful to Prof. Hilmar Meissl (Bad Nauheim), Prof. Peter Redecker (Hannover), and Prof. Manfred Ueck (Giessen) for providing us with previously unpublished photographs, to Dr. Erik Maronde, Dr. Faramarz Dehghani, Dipl. Biol. Martina Pfeffer, Dipl. Biol. Charlotte von Gall, Dipl. Biol. Bernd Neeb, Dipl. Biol. Michael Kopp, and cand. med. Karin Brednow (all Frankfurt/Main) for contributing unpublished results. The technical assistance by R. Kühn, E. Laedtke, S. Leslie, G. Müller, and I. Szasz is gratefully acknowledged. Finally, we would like to express our gratitude to Prof. Theodor H. Schiebler for his encouragement to prepare this contribution.

Our own research has been supported by the Deutsche Forschungsgemeinschaft.

Frankfurt am Main, September 1997

H.-W. Korf
C. Schomerus
J. H. Stehle

Contents

Abbreviations

ACh	Acetylcholine
AChE	Acetylcholinesterase
AP-1	Activator protein-1
CRE	Cyclic AMP response element
CREB	Cyclic AMP response element binding protein
HIOMT	Hydroxyindole-*O*-methyltransferase
ICER	Inducible cyclic AMP early repressor
IEG	Immediate early gene
LHRH	Luteinizing hormone releasing hormone
NAT	Serotonin-*N*-acetyltransferase
NE	Norepinephrine
NPY	Neuropeptide Y
OT	Oxytocin
PACAP	Pituitary adenylate cyclase-activating peptide
PCR	Polymerase chain reaction
pCREB	Phosphorylated CREB
PHI	Peptide N-terminal histidine and C-terminal isoleucine
PKA	Protein kinase A
SCG	Superior cervical ganglion
SCN	Suprachiasmatic nucleus
TPH	Tryptophan hydroxylase
TRE	Serum response element
VIP	Vasoactive intestinal peptide
VP	Vasopressin

1 Introduction

Since the origin of life, organisms have been profoundly influenced by the cyclic lighting conditions of their environment, which result in daily (diurnal) and seasonal rhythms. Plants and animals respond to rhythmic changes in the environmental lighting conditions via reactive and – even more interestingly – via anticipatory mechanisms. The anticipation of environmental light pulses requires an internal program providing a kind of "memory" about the length of the dark and light phases. To fully adapt body functions to the temporal organization of the environment this internal program needs (a) to be adjusted to environmental conditions via afferent connections and (b) to convey its signals to the body via efferent pathways. In vertebrates, including humans, these functions are carried out by a specific circuit of the brain, the photoneuroendocrine system.

An initial general concept on photoneuroendocrine systems was introduced in 1964 by Ernst Scharrer, who together with his wife, Berta Scharrer, and Wolfgang Bargmann founded the concept on neurosecretion and the discipline of neuroendocrinology. According to Ernst Scharrer, photoneuroendocrine systems serve the translation of photic stimuli into neuroendocrine responses and are distinguished from the visual system transforming light pulses into synaptic responses, and thus serving image analysis via neuronal mechanisms. Whereas the visual system is responsible for the spatial orientation, the photoneuroendocrine systems allow animals and humans to measure and keep the time. Scharrer's concept on photoneuroendocrine systems reflects an extension of his early work performed in the laboratory of Karl von Frisch in Munich and published in 1928. In the course of these studies Scharrer discovered the secretory neuron in the preoptic magnocellular nucleus of the hypothalamus in the minnow. This teleost species displays a vivid and rapid color change which was the focus of a comprehensive study performed by the scientific mentor of Scharrer, Karl von Frisch.

Von Frisch (1911) observed that shading of the head of blinded animals causes a contraction of the melanophores, whereas illumination results in an expansion of the melanin pigment. After several experimental manipulations von Frisch concluded that the light-sensitive region regulating this response must be located within the skull and suspected it to be the pineal organ that, as pointed out by Studnicka (1905), resembles a sense organ. Pinealectomy completely abolishes the melanophore response for 1 day, but after this period the light-dependent reactivity of the melanophores returns. From his experiments von Frisch concluded that the pineal organ of the minnow is a principal but not the only site of extraocular photoreception. This statement is still valid and applies for several nonmammalian species. Interestingly, the extraocular, extrapineal photoreceptor postulated by von Frisch as early as 1911 has not yet been deciphered today, although several modern experiments point to the lateral septum or the hypothalamus as possible location of this structure.

This historical excursion demonstrates that von Frisch and Scharrer can be considered as founders of photoneuroendocrinology. In a similar way of thinking, Hollwich distinguished in the retinal photoreceptive system an optic portion (comparable to the visual system of Scharrer) from an energetic portion which is important for the autonomic (unconscious) vital functions, and which is comparable to the photoneuroendocrine system of Scharrer (see Hollwich 1979, for review).

Since the original definition of photoneuroendocrine systems by Ernst Scharrer several investigations have been performed on this system (Oksche and Hartwig 1979; Korf and Oksche 1986; Korf 1994, 1996, for review), and today we know that photoneuroendocrine systems comprise three key components: (a) photoreceptor cells perceiving and transmitting environmental light pulses; (b) oscillators generating an endogenous rhythm which is independent from any environmental time cue (zeitgeber) and is called "circadian" because its period is approximately (*circa*) the length of the day (*dian*); (c) endocrine and neuroendocrine effectors receiving signals from the photoreceptors and the endogenous oscillators and translating them into a hormonal or neurohormonal response (Fig. 1; Korf 1994). This general organization appears to apply to both vertebrate and invertebrate species. The latter are becoming increasingly important models in photoneuroendocrinology, which appear very suitable to decipher the molecular basis of time keeping and time measuring systems and to identify genes that are involved in clock functions (see Hall 1995; for review).

The pineal organ (pineal gland, pineal complex, epiphysis cerebri) is an integral part of the vertebrate photoneuroendocrine system. In several nonmammalian species (e.g., lamprey, zebra fish, house sparrow) the pineal harbors a complete photoneuroendocrine system, i.e., photoreceptors, endogenous oscillators and neuroendocrine effectors. These three key components of a photoneuroendocrine system may even reside within a single cell, the photoneuroendocrine pinealocyte (see Oksche et al. 1987).

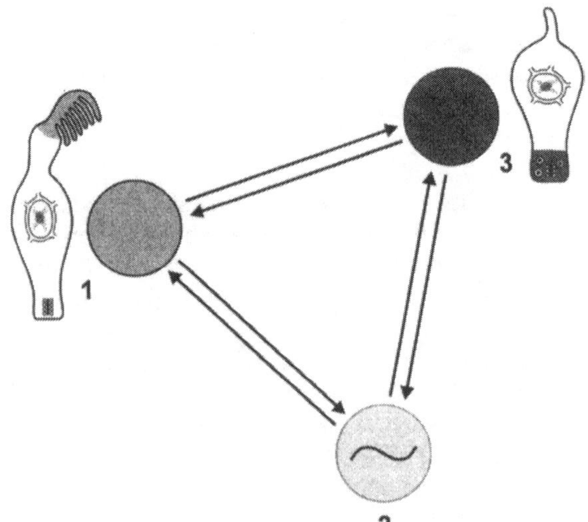

Fig. 1. Key elements of the vertebrate photoneuroendocrine system. *1*, Photoreceptor cell (located in the pineal organ or in the retina); *2*, endogenous oscillator (located in the pineal organ or the suprachiasmatic nucleus); *3*, Neuroendocrine effector (e.g., melatonin-producing pinealocyte). (Modified after Korf 1994)

The organization of both the photoneuroendocrine system and the pineal complex have in the course of evolution undergone a conspicuous transformation that indicates a high adaptive plasticity of the system (Fig. 2). The final stage of this transformation is reached in mammals, where the three key components of the photoneuroendocrine system are no longer confined to a single organ (the pineal) but are spread out into three distinct regions all of which are parts of the diencephalon. The photoreceptor cells are located in the retina. It is not yet clear, however, whether they represent rods or cones or belong to a class of photoreceptors which is distinct from these two types and remains to be characterized. The endogenous oscillator resides within the suprachiasmatic nucleus (SCN) of the hypothalamus where each single neuron appears capable of generating a circadian rhythm on its own (Welsh et al. 1995). Thus it seems that the mammalian SCN is composed of multiple circadian oscillators (Welsh et al. 1995). Very recently it has been shown that CLOCK, considered as a good candidate for a clock gene in mammals (Antoch et al. 1997; King et al. 1997; Reppert and Weaver 1997) is abundantly expressed in the SCN and in other brain areas (King et al. 1997). The mammalian pineal organ has retained its role as a major neuroendocrine effector of the photoneuroendocrine system but has lost the direct photosensitivity and the capacity to generate endogenous (circadian) oscillations. Due to this spatial separation the photoneuroendocrine system of mammals strongly depends on neuronal and endocrine pathways that interconnect its components, also in the sense of feedback loops (Fig. 3). The neuronal pathways involve circuits of both the central and the peripheral nervous system. For obvious reasons such pathways are best developed in mammals, but some portions can be traced back to very basic vertebrates, for example, the lamprey (Weigle et al. 1996).

In spite of the transformations that occurred in the course of evolution the pineal gland functions as a photoneuroendocrine transducer in all vertebrate classes. It produces and secretes melatonin, an indoleamine, in a rhythmic pattern during nighttime in response to photoperiodic stimuli and signals from endogenous circadian oscillators. This rhythm is a feature common to all vertebrate species, irrespective of whether the animals are active during the day or during the night. Another common feature is that light stimuli consistently suppress the melatonin synthesis. Melatonin can thus be considered as the neuroendocrine messenger of darkness and a timing hormone. It was isolated from the bovine pineal organ by Lerner and colleagues (1958, 1960) and identified as the substance that causes pigment aggregation in amphibian melanophores (McCord and Allen 1917; Lerner et al. 1958, 1959, 1960; see Rollag 1988). For a discussion on the functional significance of other methoxyindoles occurring in the pineal gland, see Pévet (1985).

The main biosynthetic steps were clarified soon after the discovery of melatonin (Fig. 4). Its biosynthesis starts with the uptake of circulating tryptophan into pinealocytes (Wurtman and Anton-Tay 1969) and involves 5-hydroxylation by tryptophan hydroxylase (TPH; Lovenberg et al. 1967); 5-hydroxytryptophan is transformed into serotonin (5-hydroxytryptamine) by aromatic L-amino acid decarboxylase (Snyder et al. 1965). The concentration of serotonin in the pineal gland (approximately 0.5 mM) is higher than in any other tissue except for the raphe nuclei of the midbrain (Quay 1963; Saavedra et al. 1973). The next step is the formation of N-acetylserotonin catalyzed by serotonin-N-acetyltransferase (NAT; Weissbach et al. 1960, 1961). Finally, N-acetylserotonin is O-methylated and converted into melatonin by means of the hydroxyindole-O-methyltransferase (HIOMT; Axelrod and Weissbach 1960, 1961).

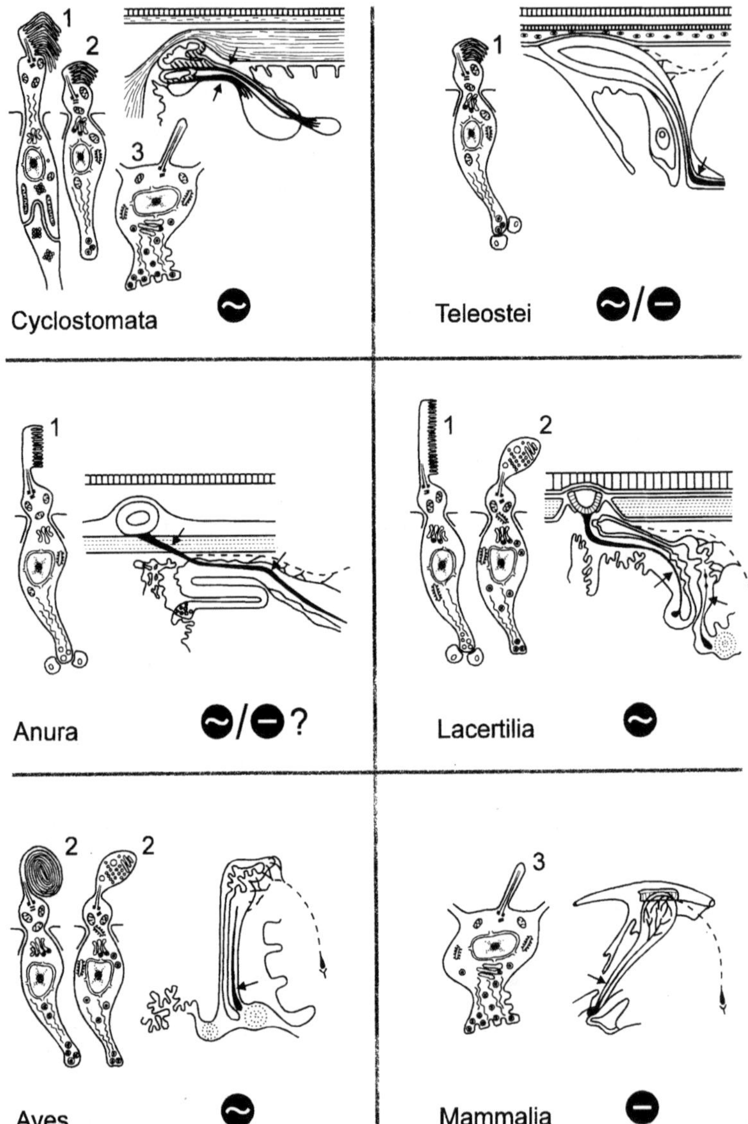

Fig. 2. Phylogenetic transformation of the pineal complex. Macroscopic appearance of the pineal organ as seen in the sagittal plane and ultrastructure of pinealocytes of the sensory line in cyclostomes, teleosts, anurans, lacertilians, birds and mammals. *Dotted lines*, noradrenergic (sympathetic) nerve fibers; *arrow*, central innervation; *1*, true pineal photoreceptor with regularly lamellated outer segment and synaptic connections with second-order neurons; *2*, modified pinealocyte with irregular outer segment or bulbous cilium; *3*, neuroendocrine pinealocyte of the mammalian type lacking an outer segment and the direct photosensitivity. In most vertebrate classes the pineal complex comprises a circadian oscillator (~); in some teleosts (e.g., rainbow trout) and in all mammals the pineal oscillator is absent (—). The noradrenergic innervation develops progressively in the course of evolution. In teleosts and anurans noradrenergic nerve fibers are only found in the capsule of the pineal. In reptiles they penetrate into the pineal; in birds and mammals they form a dense network within the pineal. (Modified after Oksche et al. 1987; Korf and Oksche 1986; Korf 1994)

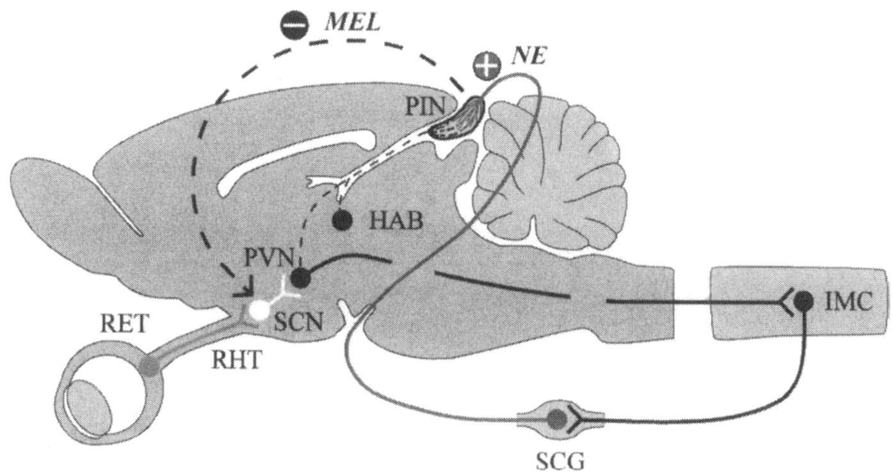

Fig. 3. The photoneuroendocrine system of the rat. Ganglion cells of the retina (*RET*) innervate the hypothalamic suprachiasmatic nucleus (*SCN*) via the retinohypothalamic tract (*RHT*). The SCN harbors the circadian oscillator. It projects to the paraventricular nucleus (*PVN*), which gives rise to two pinealopetal pathways: (a) The central innervation (*dashed lines*) reaches the pineal gland directly via its stalk; fibers from the habenular nucleus (*HAB*) join this projection. (b) The PVN forms a monosynaptic connection to the intermediolateral column (*IMC*) in the thoracic spinal cord, which gives rise to preganglionic sympathetic nerve fibers projecting to the superior cervical ganglion (*SCG*). The SCG is the origin of postganglionic nerve fibers which innervate the pineal organ (*PIN*) and release the neurotransmitter norepinephrine (*NE*) in a circadian fashion. NE activates the synthesis of melatonin (*MEL*), which feeds back to inhibit SCN activity. (Modified after Korf 1996)

According to current concepts, the highly lipophilic melatonin is not stored within the pinealocytes but released into pineal capillaries immediately after its formation. Thus the secretion of melatonin solely depends on its biosynthesis. Initially it was thought that the rhythmic changes in synthesis (and release) of melatonin are generated by daily changes in the activity of the final enzyme of the melatonin biosynthetic pathway, HIOMT. Subsequent studies, however, have shown that the large changes in synthesis and release of melatonin coincide with large changes in the production rate and availability of N-acetylserotonin and that, in the rat, the activity of the NAT is elevated 70- to 100-fold in the night (Klein and Weller 1970). Since a nocturnal increase in enzyme activity has been found in all vertebrates examined to date, NAT appears as the key regulating enzyme responsible for the large daily changes in melatonin production in the pineal gland. The magnitude of the NAT rhythm shows conspicuous variation among the species; it is most pronounced in the rat pineal gland. The activation of NAT that occurs night by night is mediated by cyclic AMP and calcium ions as second messengers.

A major breakthrough toward a molecular understanding of photoneuroendocrine systems was the recent cloning of NAT from rat and sheep pineal libraries (Borjigin et al. 1995; Coon et al. 1995; Klein et al. 1996; Roseboom et al. 1996). In the rat the time course of NAT transcription closely matches the time course of NAT activity (Borjigin et al. 1995; Roseboom et al. 1996), suggesting that the changes in NAT activity are primarily regulated by transcriptional mechanisms. However, the remarkable species-

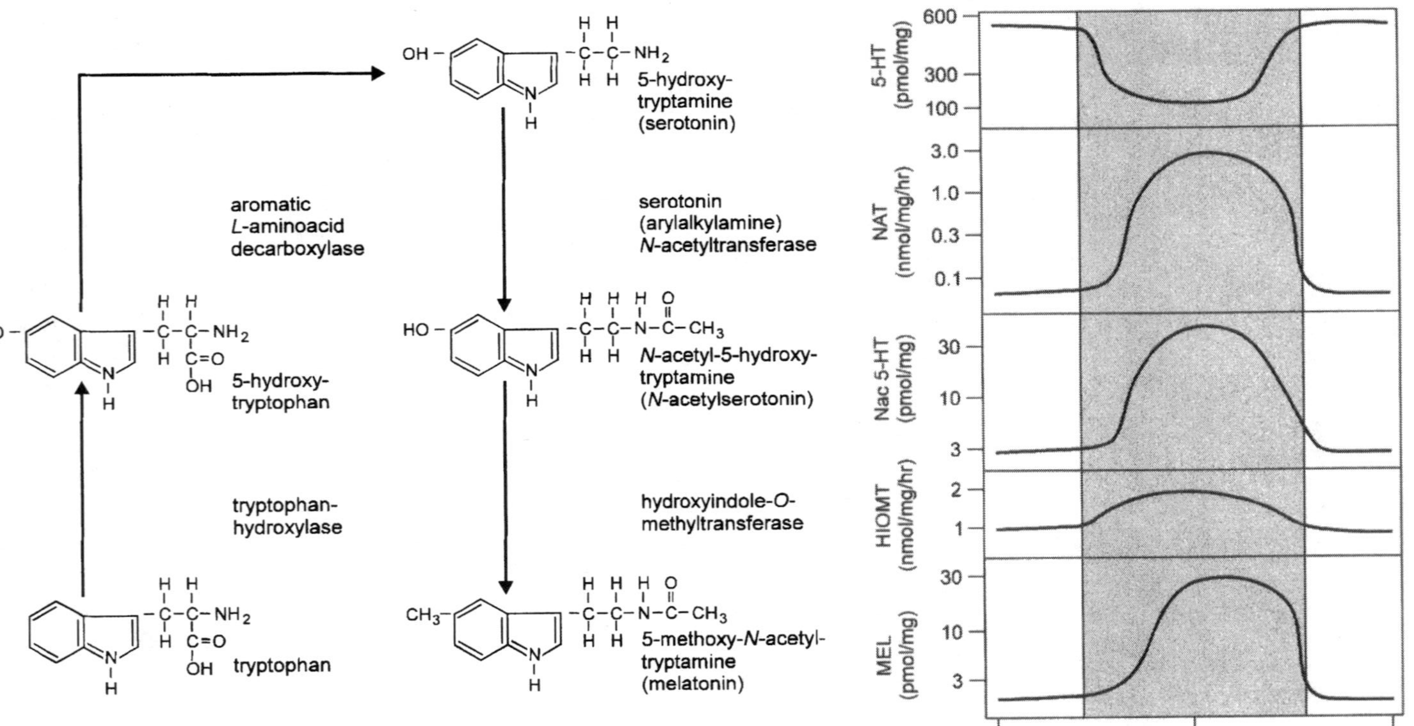

Fig. 4. *Left*, steps and enzymes of the melatonin biosynthesis; *right*, diurnal rhythms in the concentration of serotonin (*5-HT*), *N*-acetylserotonin (*Nac 5-HT*), and melatonin (*MEL*) and in the activity of the serotonin-*N*-acetyltransferase (*NAT*) and the hydroxyindole-*O*-methyltransferase (*HIOMT*) as determined in the rat pineal organ. (After Klein 1985)

to-species variation in the magnitude of the day-night NAT mRNA rhythm implies an inherent flexibility in this regulation which may also involve posttranscriptional modifications (Axelrod 1974; Klein et al. 1996). In the rat, and obviously also in other species, the transcriptional and translational upregulation, maintenance and down-regulation of NAT is essentially coupled to the cyclic AMP pathway, whose activation may be potentiated by the concomitant elevation of the intracellular calcium concentration ($[Ca^{2+}]_i$). Activation of the cAMP pathway has been shown in rat pinealocytes to cause phosphorylation of the transcription factor cyclic AMP response element (CRE) binding protein (CREB) and stimulate expression of the inducible cyclic AMP early repressor (ICER). These two transcription factors appear as a crucial link conveying activation of the second messenger system (cAMP) to the activation and inactivation of pineal gene transcription (Stehle et al. 1993, 1995; Stehle 1995; Roseboom and Klein 1995; Tamotsu et al. 1995; Foulkes et al. 1996a,b; Schomerus et al. 1996; Korf et al. 1996). In addition, other transcription factors such as Fra-2 may be involved (Baler and Klein 1995; Baler et al. 1997).

From a functional point of view, the rhythmic production of melatonin appears essential for seasonal reproduction and maternal-fetal communication (Fig. 5); it influences activity and sleep and modulates the function of the endogenous rhythm generator which plays an important role also in human physiology and pathology (sleep-wake cycle; shift work, rapid travel across several time zones; seasonal affective disorders; see Arendt 1995). These effects of melatonin appear as a part of a feedback loop connecting the suprachiasmatic nucleus as endogenous oscillator with the pineal organ as neuroendocrine effector (Fig. 3). Moreover, melatonin apparently influences the immune system (Maestroni and Conti 1991). In amphibians melatonin provokes the "primary chromatic response", i.e., the blanching reaction occurring in darkness (see Rollag 1988; Fig. 5).

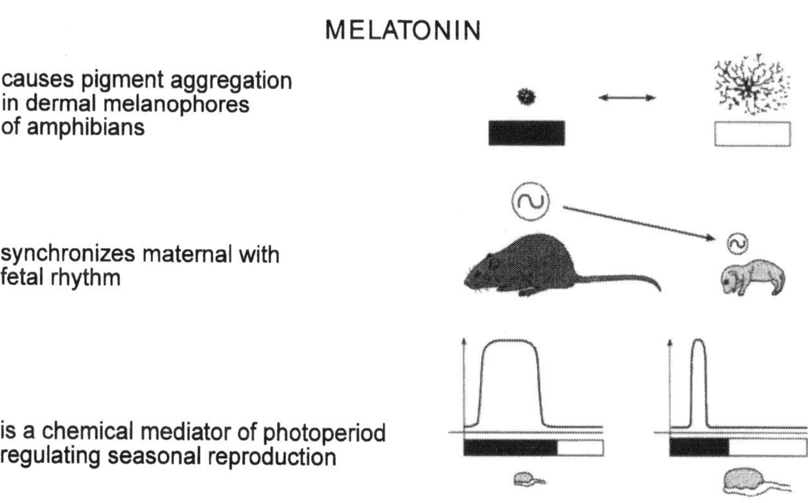

Fig. 5. Established actions of melatonin

During the past decade the targets of melatonin have been identified in various mammalian and nonmammalian species with the use of iodinated melatonin as a ligand (Vanecek et al. 1987; Weaver et al. 1989, 1993; Morgan et al. 1991a,b; Laitinen and Saavedra 1990; see Morgan et al. 1994). Although these targets vary considerably among the species, the suprachiasmatic nucleus of the hypothalamus is a site of melatonin binding in nearly all mammals investigated (see Morgan et al. 1994). In seasonal breeders the highest density of melatonin binding is found in the pars tuberalis of the adenohypophysis (Morgan et al. 1994). Melatonin binding sites are obviously more widespread in nonmammals than in mammals (Stehle 1990; Ekström and Vanecek 1992); they are found in various di- and mesencephalic brain areas primarily associated with visual functions but not in the pituitary. In amphibians displaying physiological color change mechanisms melatonin receptors are also found on dermal melanophores (see Rollag 1988; Ebisawa et al. 1994).

Molecular biological investigations performed by the group of Reppert (Ebisawa et al. 1994; Reppert et al. 1995a,b) have identified three types of melatonin receptors, referred to as Mel_{1a}, Mel_{1b}, and Mel_{1c}, and we are at the brink of understanding the signal transduction cascades that are influenced by these receptors that, as a general rule, appear to play an inhibitory role, primarily by acting upon the cAMP pathway. Recently the possibility has also been discussed that melatonin exerts its effects not only via its well-documented action upon membrane-bound receptors, but also through receptors which, as those for steroids, are located in the cell nucleus. To date, however, there is very little experimental evidence in support of the existence of such nuclear melatonin receptors.

2 Phylogenetic Development and Gross Anatomy of the Pineal Complex

Development and gross anatomy of the pineal complex have been repeatedly and comprehensively dealt with (Studnicka 1905; Bargmann 1943; Vollrath 1981; Korf and Oksche 1986; Korf 1994); thus this chapter reviews only some basic facts (Fig. 2). The pineal develops from a circumscribed area of the diencephalic roof between the habenular and posterior commissures. Interestingly, the other key components of the photoneuroendocrine system, the retina and the suprachiasmatic nucleus, are also derivatives of the diencephalon. With a few exceptions (e.g., hagfish, crocodiles), a pineal complex is present in all vertebrates investigated thus far. In many species whose pineal complex is endowed with functional photoreceptor cells the tissues overlying the pineal region display peculiar specializations; the most pronounced is the parietal foramen of the skull, which is already present in fossils, suggesting the existence of a pineal complex in Silurian and Devonian vertebrates, the ancestors of recent fish, amphibians, and lacertilians. In several recent vertebrates the pineal complex is divided into two distinct components.

In lampreys, most gnathostome, and teleost fish the pineal complex consists of a pineal organ proper (epiphysis cerebri) and a parapineal organ both of which are located inside the skull (Fig. 2). The parapineal organ of lampreys comprises a wide lumen that is reduced to a capillary space in the teleost parapineal. In all fish species the parapineal is neuronally connected to the left habenular nucleus. The pineal organ proper of lampreys and teleosts displays a wide lumen that is in open communication with the third ventricle.

In amphibians the gross anatomy of the pineal complex shows striking variation. In anurans a frontal organ located in an extracranial position in the skin can be distinguished from the intracranial pineal organ proper (Fig. 2). The frontal organ may degenerate during ontogeny (Korf et al. 1981). Obviously Urodela and Gymnophiona lack a frontal organ. The pineal organ proper of all amphibian species is a hollow structure whose lumen communicates with the third ventricle. It is in close topographical relationship with the choroid plexus of the third ventricle, the habenular and posterior commissures.

Also the reptilian pineal complex is highly variable. Apparently it is lacking in certain crocodiles (e.g., *Alligator mississippiensis*, *Crocodylus niloticus*). In other reptilian species the pineal organ proper occurs fairly regularly in typical location between the habenular and posterior commissures. The distal component of the reptilian pineal complex, the parietal eye, is located in the parietal foramen of the skull (Fig. 2). In the adult stage it persists in certain species of lizards only. The persistence obviously depends on the latitudes in which the animals live: it is higher in species living in higher latitudes. If present, the parietal eye is highly differentiated, possessing a cornea and a lens and thus resembling the lateral eyes. The parietal eye is connected to the left habenular nucleus via the parietal nerve (Korf and Wagner 1981). The pineal organ proper comprises a conspicuous lumen in lacertilian species; it is solid and glandlike in ophidians.

A similarly wide range of variation is found in the structure and appearance of the avian pineal complex which develops from a single thickening of the diencephalic roof and comprises the pineal organ proper. In passerine birds the pineal is a hollow, saclike structure, whereas in the duck and pigeon it displays a follicular shape. In sexually mature galliform birds the epiphysis is a compact, parenchymal organ.

Also in mammals the pineal complex exhibits interspecific variation in size and shape. The pineal of the opossum is a hollow evagination between the habenular and posterior commissures. In other species (e.g., hedgehog, cat, sheep, bovine, most primates including man) the pineal parenchyma forms a solid mass located between the habenular and posterior commissures (Fig. 6). The rodent pineal is composed of a superficial and deep pineal organ; these are connected by a stalk portion (Fig. 2). Obviously the size of the pineal organ is related to the metabolic activity of the organ: the pineal organ of adult mice of strain C3H which is capable of producing and secreting melatonin is twice as big as that of mice of the strain C57BL which are apparently not able to produce melatonin by virtue of a genetic defect (Fig. 7).

In all mammals the pineal is in spatial relationship with the third ventricle and the subarachnoid space. The third ventricle protrudes into the proximal portion of the pineal rather deeply, forming the pineal recess. A special feature of the mammalian pineal organ is the presence of calcareous concretions which are composed of organic and inorganic material. Their functional significance remains unknown. As shown in humans and rats, the number of pineal concretions increases with age (Vollrath 1981; Humbert and Pévet 1991), and this has been taken as a sign of gradual inactivation of

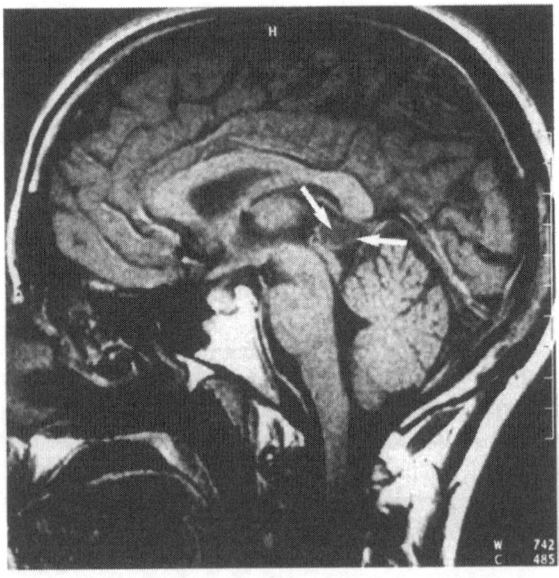

Fig. 6. Magnetic resonance tomogram showing the human brain with the pineal organ (*arrows*) in the mid-sagittal plane. (Courtesy of Prof. Dr. F. Zanella, Institut für Neuroradiologie, Frankfurt/Main)

10

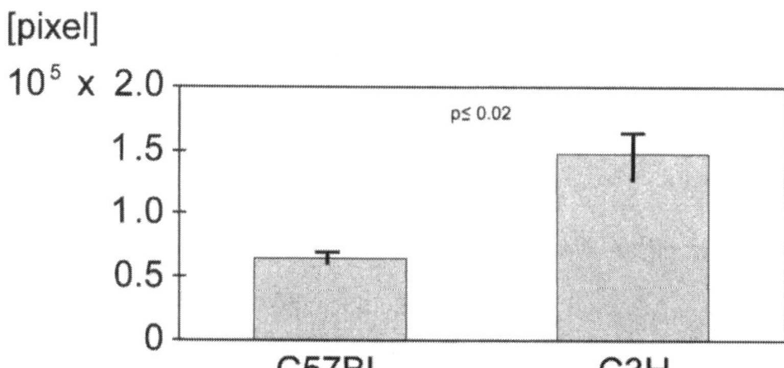

Fig. 7. Comparison of the total area (given in pixels) of the superficial pineal organ of a melatonin-producing mouse strain (C3H) and a melatonin-deficient mouse strain (C57BL). Mean values and standard deviation are calculated from pineal organs of three adult males of each strain. (K. Brednow and H.-W. Korf, unpublished)

the organ. Other investigations have, however, shown that the occurrence of these calcifications is not paralleled by a decrease in pineal metabolic activity, but may be related to secretory processes in the pineal.

As in all other vertebrate species, the primordium of the mammalian pineal organ is located between the habenular and posterior commissures. The human pineal organ appears in embryos of 6–8 mm total length. In early ontogenetic stages an anterior lobe can be distinguished from a posterior one; the two are separated by a connective tissue septum. In the course of further development these lobes fuse and form the pineal body (Møllgard and Møller 1973; Møller 1978).

Noteworthy is the intimate relationship that exists between the habenular nuclei and the pineal complex in all vertebrate species. These areas are intimately connected by neuronal pathways. Moreover, the cytoarchitectural and histochemical features of the teleost parapineal organ are strikingly similar to those of the habenular nucleus (Korf 1974). In rodents the medial habenular nucleus comprises not only neurons but also pinealocytes (Korf et al. 1986a,b, 1990) and is thus considered as an intermediate area between the brain and the pineal organ.

3 Pineal Cell Biology and Innervation

3.1
General Aspects

In comparative terms, pinealocytes, supporting (glial) cells, and neurons form the pineal parenchyma which is separated from the adjacent connective tissue layer (capsule) and the capillaries by a basal lamina. Pinealocytes display striking variation among different classes of vertebrates. Based on structural and ultrastructural criteria they have been divided into three main categories: true pineal photoreceptors, modified pineal photoreceptors, and pinealocytes sensu stricto (Oksche 1965, 1971; Collin 1971; Collin and Oksche 1981; Korf and Oksche 1986; Korf 1994; Figs. 2, 8). Since the pioneering work of Dodt and Heerd (1962) the direct light sensitivity of true pineal photoreceptors has been firmly established. The fact that modified photoreceptor cells are also capable of perceiving light stimuli has been known since Deguchi's discovery in 1981 that, in the chicken, melatonin production is regulated by light perceived in the pineal organ. The direct light sensitivity has been lost in the mammalian pinealocyte. Nevertheless, all pinealocyte types appear to be closely related and can be classified as cells of the receptor line. This intimate relationship has been substantiated by immunocytochemical and biochemical studies showing that neuroendocrine pinealocytes of mammals express "photoreceptor-specific" molecules which otherwise are synthesized only by functional (retinal and pineal) photoreceptor cells (Korf et al. 1985a,b, 1986a,b, 1992; Huang et al. 1992; Kramm et al. 1993; Figs. 9, 10). As holds true for primary sensory cells, all types of pinealocytes belong to the neuronal cell lineage because they express a variety of markers typical of neurons and neuroendocrine cells (neurofilament, synaptobrevin). Pinealocytes can thus be considered as specialized neurons.

Pineal photoreceptors and intrapineal second-order neurons are gradually reduced in the course of evolution. In mammals and man the parenchyma of the pineal gland consists of pinealocytes sensu stricto. Concomitantly with the regression of true photoreceptor cells the pinealofugal innervation is gradually reduced in mammals. Conversely, the autonomic innervation develops progressively in the course of evolution. The autonomic (sympathetic innervation) of the pineal is most prominent and essential for the photoneural regulation of the melatonin biosynthesis in mammals. In most mammalian species the postganglionic sympathetic nerve fibers reach the pineal organ via a special, bilaterally arranged nerve running in the tentorium cerebelli, the conarian nerve. Autonomic nerve fibers also innervate the avian pineal organ, but they appear functionally less important in birds than in mammals. Autonomic nerve fibers are scarce or even absent in the pineal region of poikilothermic vertebrates such as fishes and amphibians.

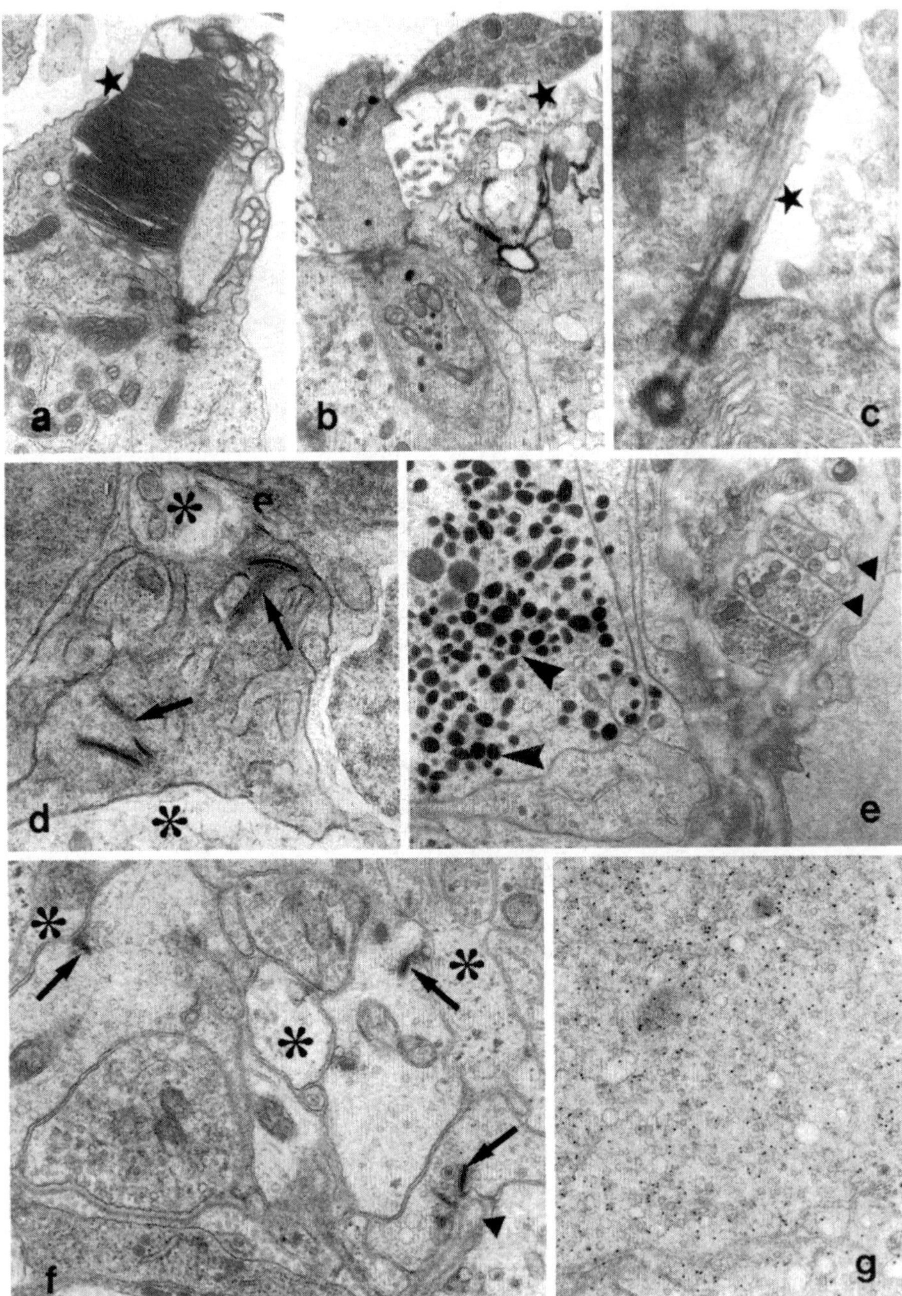

Fig. 8. Ultrastructure of the receptor (**a–c**) and effector (**d–g**) poles of different types of pinealocytes.
a True pineal photoreceptor of rainbow trout with outer segment (*star*) comprising several, regularly
arranged disks, cilium of the 9+0 type and inner segment. ×12,500 (H. Blank, B. Müller and H.-W.
Korf, unpublished) **b** Modified pineal photoreceptor of the lizard, *Lacerta agilis*, with bulbous cilium
(*star*). ×7800 (Courtesy of A. Oksche and H. Kirschstein, unpublished; see Oksche and Kirschstein

3.2
Cell Biology of True Pineal Photoreceptors, Modified Pineal Photoreceptors, and Pinealocytes of the Mammalian Type

This section deals with the three principal types of pinealocytes, the true and modified pineal photoreceptors and the pinealocytes sensu stricto. Since these types show a characteristic distribution among the different classes of vertebrates, this section is arranged according to animal classes.

3.2.1
Anamniotes

The principal cell type of the pineal complex of anamniotic vertebrates is the true or typical pineal photoreceptor which bears an outer segment that protrudes into the pineal lumen and consists of numerous disks produced by successive basoapical invaginations of the plasma membrane (Figs. 2, 8a). Depending on the species the number of outer segment disks varies between 10 and 300. The outer segment is connected to the inner segment via a cilium of the 9×2+0 type. Opposite to the outer segment, the pineal photoreceptor gives rise to a basal process originating from the perikaryon and contributing to prominent intrapineal neuropil formations. Its enlarged terminals contain numerous electron lucent synaptic vesicles intermingled with synaptic ribbons and scattered dense core vesicles (Figs. 2, 8d). Via these terminals the true pineal photoreceptors establish synapses with intrapineal second-order neurons. Adjacent photoreceptor cells are connected via gap junctions, suggesting that they are electrically coupled (see Ekström and Meissl 1997).

By means of immunocytochemical and biochemical investigations it has been shown that pineal photoreceptors contain molecules of the phototransduction cascade which are very closely related to or even identical with those expressed by retinal photoreceptors (Fig. 9). Thus immunoreaction for rod-opsin, the proteinous component of the rod visual pigment rhodopsin has been found in the outer segments of many pineal photoreceptors in lamprey (Tamotsu et al. 1990), teleosts (Vigh-Teich-

1968) c Cilium (*star*) of a mouse pinealocyte. ×24,000 (J. Timpe and H.-W. Korf, unpublished) d Synaptic ribbons (*arrows*) accompanied by clear synaptic vesicles in a basal process of a true pineal photoreceptor of the minnow, *Phoxinus phoxinus; asterisks,* postsynaptic dendrites. ×12,000 (Courtesy of A. Oksche and H. Kirschstein, unpublished; see Oksche and Kirschstein 1967, 1971) e Basal processes of modified pineal photoreceptors in the lizard, *Lacerta agilis,* containing numerous large dense-core granules (*arrowheads*); *triangles,* small bundle of noradrenergic nerve fibers that comprise numerous small granules with an eccentric dense core in addition to some larger dense-core vesicles and that are surrounded by the basal lamina. Note fenestrated endothelium of a pineal capillary adjacent to the nerve bundle. ×18500 (Courtesy of A. Oksche and H. Kirschstein, unpublished; see Oksche and Kirschstein 1968). f Basal processes of rat pinealocytes containing synaptic ribbons (*arrows*) accompanied by clear synaptic vesicles and isolated dense-core granules. The ribbons are facing either other processes (*asterisks*) or the basal lamina (*triangle*). ×24,000 g Immunocytochemical demonstration of glutamate in the basal process of a gerbil pinealocyte (immunogold method). Numerous gold particles are associated with clear synaptic vesicles. ×25,500 (Courtesy of P. Redecker)

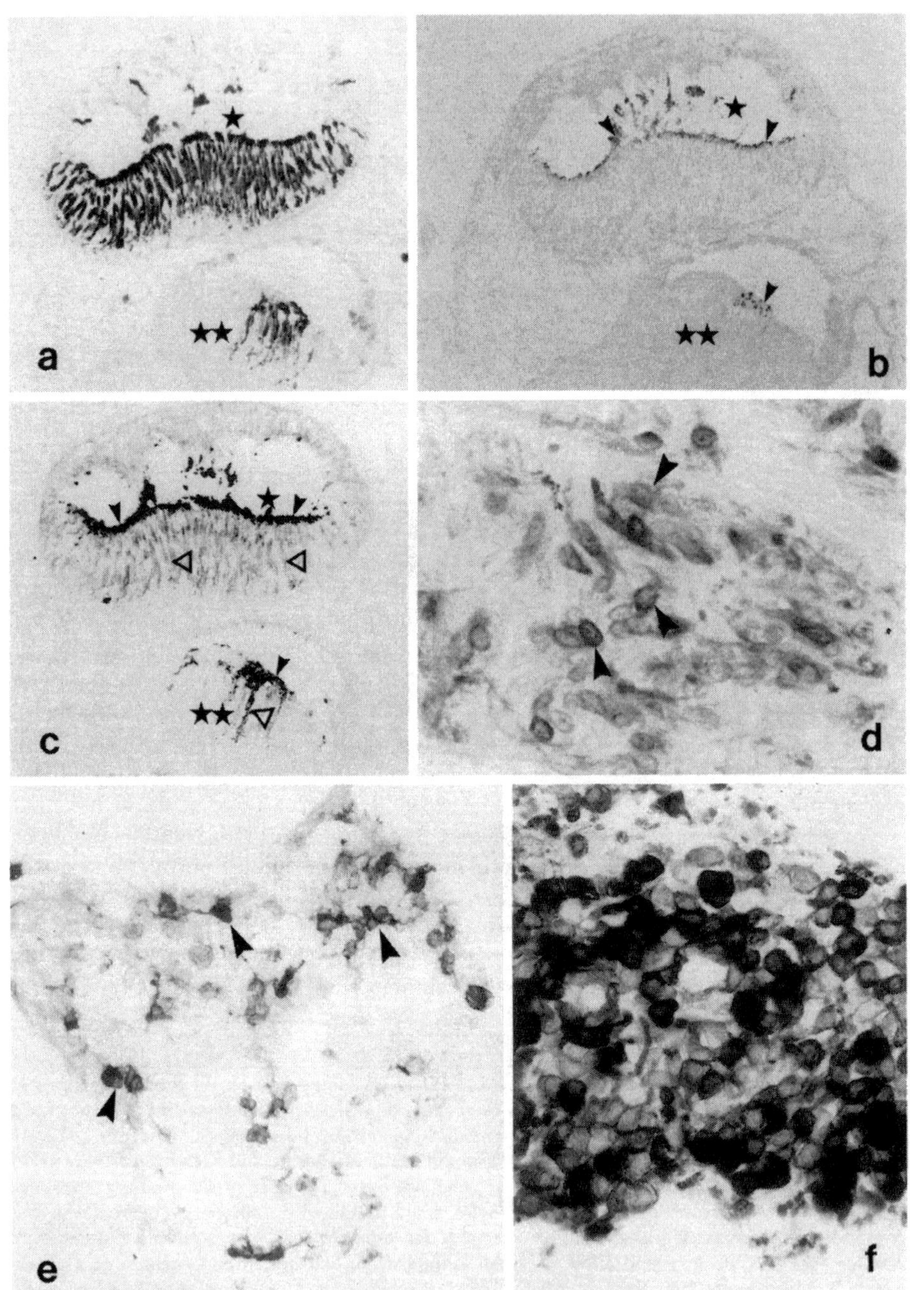

Fig. 9a–f. Immunocytochemical demonstration of photoreceptor-specific proteins in different types of pinealocytes. a–c Sagittal sections of the pineal (*star*) and parapineal (*double star*) organs of the lamprey, *Lampetra japonica*. a S-antigen immunoreaction distributed throughout all compartments of the pineal photoreceptors. b α-Transducin immunoreaction labeling pineal outer segments (*arrowheads*) protruding into the lumen. c Rod-opsin immunoreaction present in outer segments (*arrow-*

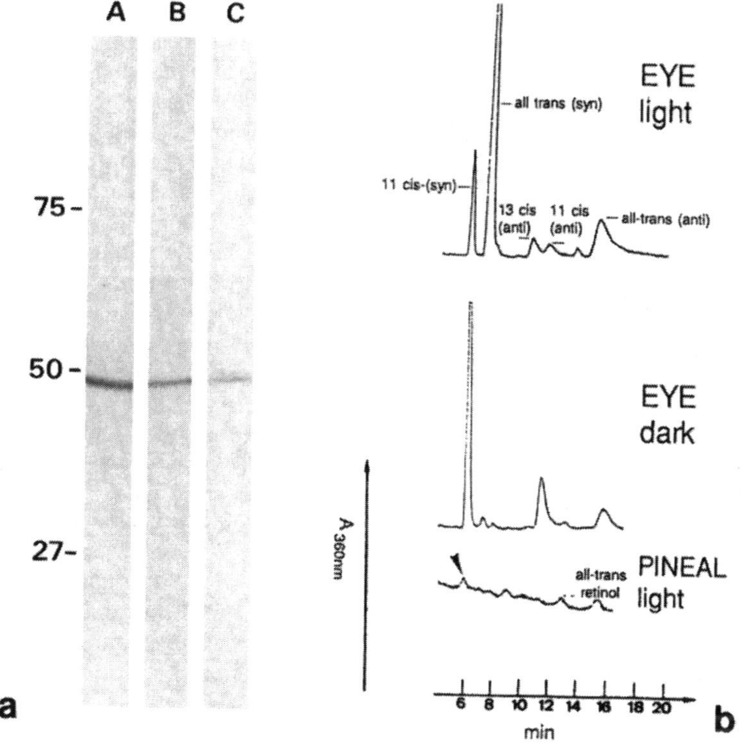

Fig. 10. a Immunoblot showing a single protein band of approximately 48 kDa in retina (*A*), superficial pineal organ (*B*) and deep pineal organ/habenular region (*C*) of the mouse which reacts with the S-antigen antiserum. (From Korf et al. 1990). b Representative HPLC chromatograms of lipid extracts from light- and dark-adapted eyes and light-adapted pineal organs of pigmented mice. In the extracts of light- and dark-adapted eyes peaks are found which represent *11-cis* (syn), *11-cis* (anti), *all-trans* (syn), all-*trans* (anti), and 13-*cis* (anti) retinaloximes. Such peaks are missing in the extract of the light-adapted pineal organs; only *all-trans* retinol can be detected. The *arrowhead* marks a small peak that is close to that of *11-cis* (syn) retinaloxime, but differs from the latter by both retention time and spectrum. (From Kramm et al. 1993)

mann et al. 1982), frogs (Vigh-Teichmann and Vigh 1990) and certain reptiles (Vigh et al. 1982). In the pineal of the clawed toad and the agamid lizard (*Uromastix hardwicki*) very few or no rod-opsin immunoreactive pineal outer segments are found (Hafeez et al. 1987; Korf et al. 1989). These species, however, have pineal photoreceptors that react with antibodies raised against chicken cone-opsin. Cone-opsin immunoreactive outer segments are also found in ranid frogs, but they are less frequent than the rod-opsin immunoreactive outer segments.

heads) and basal processes (*open triangles*). **a–c** ×70 **d** S-antigen immunoreaction in modified pineal photoreceptors (*arrowheads*) of the duck. ×270 **e** Rod-opsin immunoreaction labels a population of mouse pinealocytes (*arrowheads*). ×115 **f** S-antigen immunoreaction in gerbil pinealocytes. Note that the intensity varies on a cell-to-cell basis. ×460

Taken together, these findings indicate that multiple types of true pineal photoreceptors exist; some are closely related to rods whereas others appear conelike. Microspectrographic and electrophysiological experiments have also revealed the presence of multiple photopigments (Solessio and Engbretson 1993; Meissl and Ekström 1993; Ekström and Meissl 1997). To date it has not yet been investigated whether a subpopulation of true pineal photoreceptors contain the pineal-specific photopigment pinopsin, which has been cloned from the chicken pineal organ. This novel photopigment differs from all known retinal photopigments but has 45% homology with the opsin of blue cones (Okano et al. 1994; Max et al. 1995). For further and more comprehensive discussion on pineal photopigments and neurophysiological properties of true pineal photoreceptors, see Ekström and Meissl (1997). Many true pineal photoreceptors share additional molecular features with their retinal counterparts; they express immunoreactive α-transducin (van Veen et al. 1986), S-antigen (Korf et al. 1985b, 1986a,b) and recoverin (Korf et al. 1992).

The concept that true pineal photoreceptors belong to the neuronal cell lineage is supported by the demonstration of neuronal markers in these cells (neuron-specific enolase: Oksche et al. 1987; neurofilament 200-kDa: Blank et al. 1997; Fig. 11). Because of its high selectivity and specificity, the finding of immunoreactive neurofilament provides good arguments for the neuronal nature of true pineal photoreceptors. Interestingly, also retinal photoreceptors have been shown to contain this immunoreaction (Blank et al. 1997). In both retinal and pineal photoreceptors the immunoreaction is confined to the axoneme which connects the outer with the inner segment. These observations suggest that neurofilaments form a part of the photoreceptor cytoskeleton.

The neurotransmitter employed by pineal photoreceptors has not yet been precisely identified. Electrophysiological studies in frogs suggest that the pineal photoreceptors – as their retinal counterparts – utilize excitatory amino acids, i.e., glutamate or aspartate as transmitter (Meissl and George 1984a,b). In frogs and reptiles glutamate and aspartate have been shown in pineal photoreceptors of the pineal organ and parietal eye, respectively, by means of immunocytochemical techniques (Vigh et al. 1995, 1997). These findings conform to earlier biochemical results showing that glutamate and aspartate are present in the trout (Meissl et al. 1978) and goldfish pineal (McNulty et al. 1988).

Synaptic and neuronal mechanisms are one mode of action how the pineal organ of poikilothermic vertebrates translates the environmental lighting conditions. On the other hand, this organ generates a neuroendocrine message, melatonin, in response to the environmental lighting conditions. Both the neuronal and the neuroendocrine response are inhibited by light. In view of these dualistic effector mechanisms it is of interest to determine whether the true pineal photoreceptors in the anamniotic pineal complex are also able to produce melatonin. Investigations performed by Falcon, Collin and coworkers have suggested that true pineal photoreceptors are indeed capable of melatonin biosynthesis (Guerlotte et al. 1986). This notion has been supported by the immunocytochemical demonstration of HIOMT in virtually all pineal photoreceptors of teleosts (Falcon et al. 1994). On the other hand, immunocytochemical demonstration of serotonin, the precursor of melatonin, has unraveled conspicuous differences between various types of pinealocytes (see Ekström and Meissl 1990; Tamotsu et al. 1990; Fig. 12).

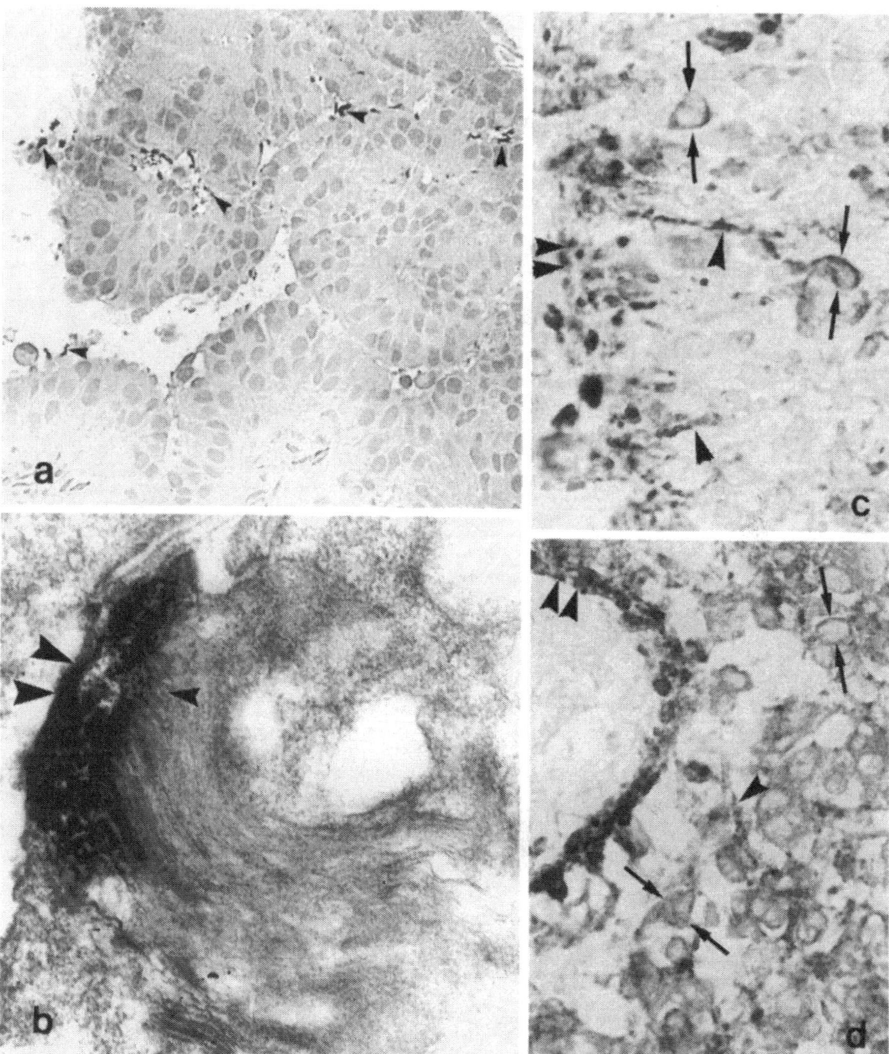

Fig. 11a–d. Neuronal markers in pinealocytes. **a–c** Immunoreaction for neurofilament 200-kDa (NF-200). **a** Immunoreactive NF-200 labels dotlike structures (*arrowheads*) protruding into the lumen of the rainbow trout pineal organ. Semithin sagittal section. ×300 (H. Blank, B. Müller, and H.-W. Korf, unpublished) **b** Immunoelectron microscopy with the preembedding technique revealed that the dotlike structures in the trout pineal organ represent NF-200 immunoreaction in the axoneme (*arrowheads*) of pineal photoreceptors which arises from the connecting cilium of the outer segment. ×32,000 (H. Blank, B. Müller, and H.-W. Korf, unpublished) **c** Immunoreactive NF-200 in human pinealocytes. **d** Immunoreactive synaptophysin in human pinealocytes. *Arrows*, immunoreactive perikarya; *arrowheads*, immunoreactive processes of pinealocytes; *double arrowheads*, strongly immunoreactive dots corresponding to endfeet of pinealocyte processes at the border between pineal follicles and connective tissue. **c,d** ×490 (For details, see Huang et al. 1992)

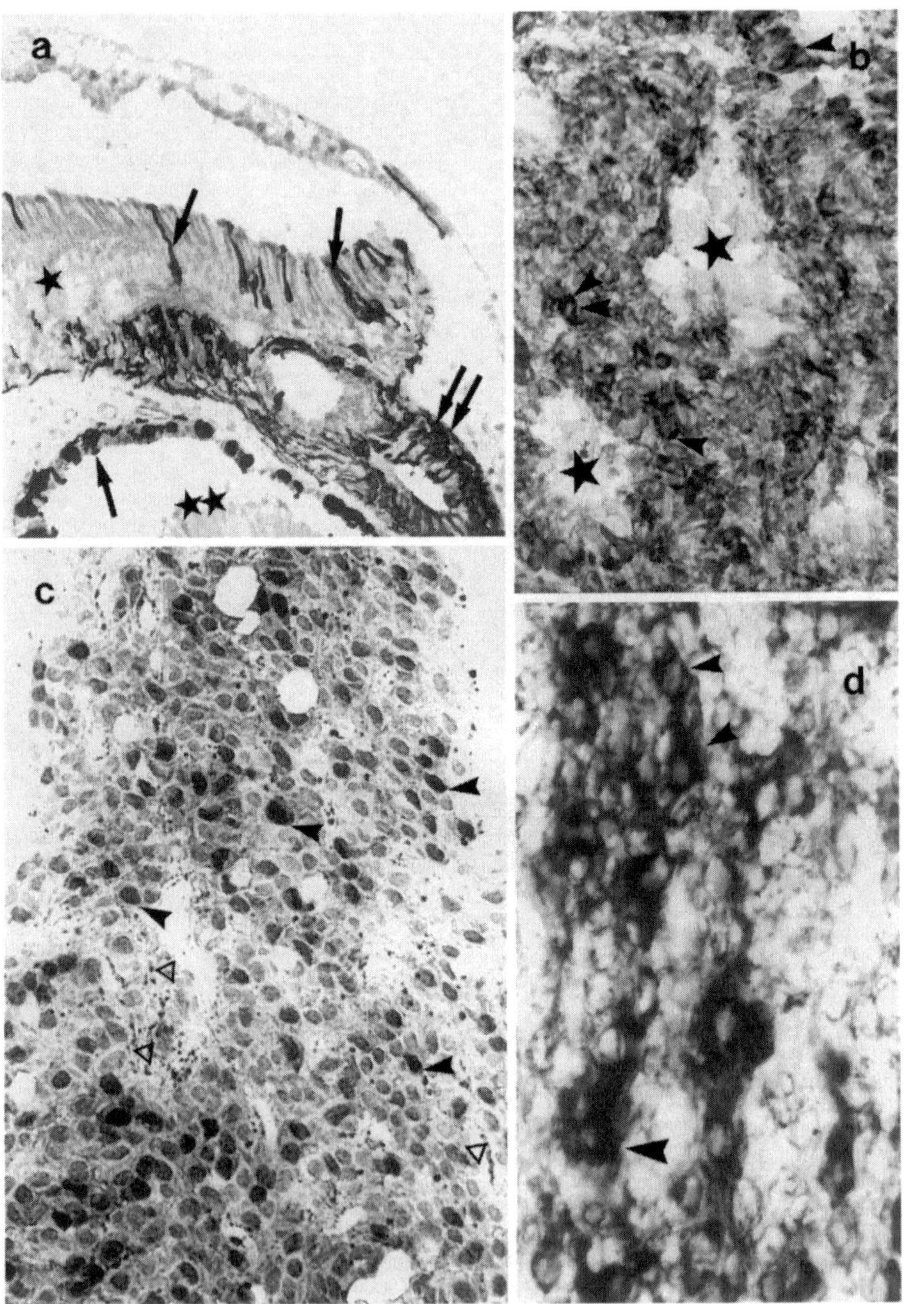

Fig. 12. Immunocytochemical demonstration of serotonin, the precursor of melatonin, (**a–c**) and of hydroxyindole-*O*-methyltransferase (HIOMT) (**d**). **a** Sagittal section of the pineal (*star*) and parapineal (*double stars*) organ of the lamprey, *Lampetra japonica*. The serotonin immunoreaction labels modified pineal photoreceptors scattered in the pineal end vesicle and the roof of the parapineal organ (*arrowheads*) and pinealocytes densely arranged in the atrium of the pineal (*double arrowheads*). ×180 (For

In the lamprey, *Lampetra japonica*, the pineal photoreceptors which bear long rod-opsin immunoreactive outer segments have been shown to lack serotonin immunoreactivity that has, however, been demonstrated in two other types of pinealocytes (Tamotsu et al. 1990). One of these has a short rod-opsin immunoreactive outer segment and displays features of the modified photoreceptor cell; the other cell type is located in the proximal portion of the pineal, the atrium. It lacks an outer segment and contains a weak S-antigen immunoreaction, thus resembling the pinealocyte sensu stricto found in the mammalian pineal organ. These results speak in favor of cell-to-cell differences in the indoleamine metabolism. They fully conform to earlier electron microscopic and fluorometric results (Meiniel 1981) and support the notion that multiple types of pinealocytes are present at an early evolutionary state. Also in the pineal organ of the pike true pineal photoreceptors have been found to coexist with modified photoreceptor cells (Falcon 1984). These findings are relevant for considerations on the cytoevolution of pinealocytes. One idea is that mammalian pinealocytes represent the final stage of a gradual transformation of true pineal photoreceptors characteristic of anamniotes. The occurrence of all three types of pinealocytes (i.e., true and modified photoreceptors as well as pinealocytes of the mammalian type) in the pineal organ of the lamprey (a most basic vertebrate) indicates that they have evolved in parallel.

3.2.2
Reptiles (Lacertilians and Ophidians)

The pineal complex of lizards contains both true and modified photoreceptor cells (Oksche and Kirschstein 1968). The best developed true pineal photoreceptors are found in the parietal eye (Fig. 6; see Vollrath 1981, for review and references). Modified photoreceptors are endowed with rudimentary outer segments (Fig. 8b), and their basal processes contain numerous electron-dense granules (Fig. 8e). The solid pineal organ of ophidians comprises pinealocytes which differ from true and modified photoreceptors. They lack outer and inner segments but contain cilia of the 9+0 type and synaptic ribbons. Their basal processes are filled with an abundant amount of dense-core granules. Fluorescence microscopy has shown that pinealocytes of *Natrix natrix* contain serotonin (Quay et al. 1968). Interestingly, antibodies against rod-opsin, cone-opsin, and S-antigen did not stain any pinealocyte in the pineal organ of *Naja naja*, *Vipera russeli*, or *Echis carinatus* although they labeled photoreceptors in the retina of all three species (Hafeez et al. 1995). Also in the garter snake no pinealocytes could be stained with antibodies against rod-opsin, and only two out of four monoclonal S-antigen antibodies elicited a weak immunoreaction (Kalsow et al. 1991). These results suggest that the ophidian pinealocytes resemble a type of pinealocyte which

---

details, see Tamotsu et al. 1990) **b** Serotonin immunoreaction in modified pineal photoreceptors of the duck (*arrowheads*). *Stars,* lumen of two pineal follicles. ×165 **c** Sagittal section of the gerbil pineal organ. Strongly serotonin-immunoreactive pinealocytes (*arrowheads*) can be distinguished from weakly labeled cells. Dotlike immunoreactive structures (*open triangles*) correspond to sympathetic nerve fibers that have taken up serotonin. ×270 **d** Sagittal section of the bovine pineal organ showing HIOMT immunoreaction in several (*arrowheads*) but not all pinealocytes. ×560

predominates in the pineal organ of primates and man and is also immunonegative for rod-opsin and S-antigen (Huang et al. 1992).

3.2.3
Birds

The predominant and most conspicuous cell type of the avian pineal organ is the modified pineal photoreceptor (Fig. 2). The outer segment of these cells is less regular than that of true pineal photoreceptors. Some modified photoreceptors have only a bulbous cilium lacking membrane disks. The basal process of the cells contains synaptic ribbons intermingled with clear vesicles and dense-core granules. The number of the latter varies considerably on a species-to-species basis, but in general these granules are much more numerous in modified than in true pineal photoreceptors. The dense-core granules are taken as an indication of a high secretory activity of the modified photoreceptor cells, although the content of these granules has not been clarified. According to current concepts, melatonin is not stored within the pinealocyte but is released immediately after its formation. Thus it may be speculated that these granules contain a neuroactive substance that differs from melatonin. Since they closely resemble the granules of peptidergic neurons, one may assume that they harbor a peptide that is coreleased with melatonin. The basal processes of modified photoreceptors terminate adjacent to the basal lamina or are apposed to basal processes of other modified photoreceptors. Obviously the basal processes of modified photoreceptors do not form synaptic contacts with intrapineal neurons. Their number is somewhat reduced in adult birds, with the exception of passerine species whose pineal organ comprises several acetylcholinesterase (AChE) positive nerve cells and a well-developed pineal tract (Ueck and Kobayashi 1972; Korf et al. 1982).

As shown by immunocytochemistry, modified photoreceptors of the avian pineal organ contain photoreceptor-specific proteins, but, again, conspicuous species differences are evident. Several rod-opsin immunoreactive outer segment remnants or whole cells are found in the pineal of the pigeon (Vigh et al. 1982) and Pekin duck (Korf and Vigh-Teichmann 1984). In contrast, there are only very few rod-opsin positive elements in the Japanese quail (Foster et al. 1987) and chicken (Sato et al. 1990; Korf 1994). Most probably the latter finding can be explained by the fact that in the chicken pineal organ pinopsin is the predominant photopigment (Okano et al. 1994; Max et al. 1995; Hirunagi et al. 1997), and that the rod-opsin antibody used did not cross-react with pinopsin, which has 45% homology with the opsin of blue cones in the retina. Interestingly, also the S-antigen immunoreaction, the marker most widely distributed in all types of vertebrate pinealocytes, is absent from the chicken pineal organ, although the S-antigen immunoreaction is expressed by photoreceptors in the chicken retina. These results suggest that the arrestin (S-antigen) molecule of the pinopsin phototransduction cascade differs from the arrestin of the rhodopsin phototransduction cascade. Many strongly S-antigen immunoreactive modified pineal photoreceptors are found in the quail, duck, and pigeon (Fig. 9d). These species have also been shown to contain α-transducin, recoverin and interstitial retinol binding protein immunoreactive cells (van Veen et al. 1986; Foster et al. 1987; Korf et al. 1992).

The immunocytochemical results conform to those of studies showing the direct light sensitivity of the chicken pineal organ (Deguchi 1981; Takahashi et al. 1989; Zatz et al. 1988) and the presence of 11-*cis* retinal, the prosthetic group of any vertebrate photopigment, and its light-dependent stereoisomerization in the quail pineal organ (Foster et al. 1989a). As shown by in vitro experiments, light stimuli elicit an acute inhibition of the melatonin biosynthesis in cultured chicken pineal organs. Thus the directly light-sensitive cells in the chicken pineal organ appear to translate the photic stimulus into a neuroendocrine response and can be classified as a photoneuroendocrine cell. The capacity of the modified photoreceptor cells to produce melatonin has been corroborated by immunocytochemical demonstration of serotonin (Fig. 12b) and HIOMT in these cells (Voisin et al. 1988; Bernard et al. 1991). Moreover, in the chicken pineal some multipolar cells (interfollicular cells) display immunoreactive HIOMT.

Transplantation experiments with the house sparrow (Zimmerman and Menaker 1979) have shown that the avian pineal organ also contains an endogenous oscillator. This has been confirmed by in vitro experiments showing that the melatonin rhythm persists for several cycles when isolated fragments of the chicken pineal are kept in constant darkness; illumination of these preparations causes an entrainment of the oscillator (Robertson and Takahashi 1988; Zatz et al. 1988). These findings raise the interesting hypothesis that the key processes of the photoneuroendocrine system, namely photoreception, endogenous oscillation, and melatonin production, are accomplished in a single cell, the photoneuroendocrine pinealocyte.

It should be noted, however, that considerable differences exist in the organization of the photoneuroendocrine system among different avian species. In contrast to the sparrow, pinealectomy only occasionally abolishes the free-running circadian activity rhythms in the European starling (Gwinner 1990). Moreover, it is not clear to what extent the sympathetic innervation of the pineal contributes to the circadian regulation of the avian pineal. The sympathetic innervation which appears to originate from the superior cervical ganglion (SCG; Hedlund and Nalbandov 1969) is quite prominent in most avian species. In vivo analyses by Cassone and Menaker (1983) have shown that, in contrast to normal animals, chicken whose superior cervical ganglia were removed could not sustain persistent rhythms in constant darkness. These results support the notion that the sympathetic innervation is involved in regulation of the pineal's inherent rhythmicity.

3.2.4
Mammals

Pinealocytes sensu stricto (Fig. 2) form the main cellular component of the mammalian pineal organ. They appear as mono-, bi-, and multipolar cells (Bargmann 1943). In some species, for example, the hamster and the opossum, certain pinealocytes are directly exposed to the cerebrospinal fluid (CSF) and can be classified as CSF-contacting pinealocytes (Hewing 1978; Welsh 1983; Korf et al. 1986a). Mammalian pinealocytes lack outer and inner segments but contain cilia of the 9+0 type, presumably representing remnants of the receptor pole (Figs. 2, 8c). Although one type of pinealocyte has been shown to possess long axonlike processes leaving the pineal organ and penetrating into the brain (Korf et al. 1986b, 1990), most pinealocytes bear processes terminating within the pineal parenchyma at the basal lamina of the perivascular space. These processes contain synaptic ribbons and a varying number of clear and dense core vesicles (Fig. 8f).

There is general agreement that mammalian pinealocytes produce melatonin. However, it is not clear whether melatonin biosynthesis occurs in all or only in certain cells. Immunocytochemical demonstration of melatonin has revealed conflicting results and is considered unreliable by most immunocytochemists. Immunocytochemical demonstration of serotonin, the precursor of melatonin, has shown an even distribution in the mouse pineal organ, but in the gerbil pineal the serotonin immunoreactivity displays a conspicuous cell-to-cell variation (Fig. 12c). In the bovine not all but only certain pinealocytes are labeled with an antibody against HIOMT (Fig. 12d). After the successful cloning of NAT, some novel antibodies became available. Preliminary findings obtained with these new tools suggest that – in the sheep – the majority of pinealocytes express NAT; the intensity of the immunoreaction, however, varies considerably on a cell-to-cell basis.

The close phylogenetic relationship between mammalian pinealocytes and pineal photoreceptors is illustrated by the fact that mammalian pinealocytes display immunoreactions for photoreceptor-specific proteins, such as rod-opsin (Korf et al. 1985a; Huang et al. 1992), S-antigen (Korf et al. 1985b, 1986a,b, 1990), and recoverin (Korf et al. 1992; Schomerus et al. 1994). The number of immunoreactive cells varies with the species and the antibodies applied. In the rodent and cat pineal organ the majority of pinealocytes are S-antigen immunoreactive (Fig. 10f), whereas in the human pineal organ, the S-antigen immunoreactive cells make up only 5%–10% of the total population. The number of rod-opsin immunoreactive pinealocytes is constantly smaller than that of the S-antigen immunoreactive cells. No rod-opsin immunoreactive pinealocyte is found in the pineal organ of adult albino rodents. In humans a small subpopulation of pinealocytes (3%–5%) bind the rod-opsin antibody. Approximately 25%–30% of the pinealocytes are rod-opsin immunoreactive in adult pigmented mice (wild type and C57BL; Fig. 10e). The most interesting pattern of immunoreactions for "photoreceptor-specific" proteins has been observed in the "blind" mole rat, *Cryptomys damarensis*, which has a characteristic patch of white hairs in the parietal region of the skull. In this species approximately 50% of pinealocytes are rod-opsin immunoreactive. A considerable number of cells also displays immunoreactive α-transducin, which has never been shown with certainty in the pineal organ of adult individuals of any other mammalian species (Korf and Wicht 1992; Schomerus et al. 1994).

Immunochemical and in situ hybridization histochemical investigations have provided evidence that the described immunoreactions are indeed elicited by the authentic proteins of the phototransduction cascade. Three bands of approximately 40, 75, and 110 kDa have been found to bind the rod-opsin antibody in immunoblots of the pineal organ of the pigmented mouse (Kramm et al. 1993), and corresponding bands have been found in the mouse retina. Moreover, the S-antigen and recoverin antibodies label bands of identical molecular mass in the retina and the pineal organ of various mammalian species (Fig. 10; see Korf 1994, for review). Also, the in situ hybridization has revealed S-antigen expression in the retina and pineal organ (Korf and Wicht 1991).

The functional significance of these proteins in the mammalian pineal organ remains to be established. It is generally accepted that mammalian pinealocytes have lost the direct light sensitivity. This concept is confirmed by studies showing that the mammalian pineal organ does not contain 11-*cis* retinal, the prosthetic group essential for a functioning photopigment (Foster et al. 1989b; Kramm et al. 1993; Fig. 10b). Interestingly, molecular biological studies have revealed surprising similarities between the rhodopsin-based phototransduction cascade and the transmembrane signaling system conveying β-adrenergic stimuli. In both systems agonists induce conformational changes in the receptor protein and turn the receptor into a substrate for a specific kinase (β-receptor kinase or rhodopsin kinase, respectively). This phosphorylation then triggers the binding of a specific arresting protein (S-antigen or β-arrestin, respectively), which disrupts the coupling between the receptor and the G protein. A high homology has been found between the rod-opsin molecule and the β-adrenergic receptor (Dohlman et al. 1987). β-Receptor kinase is also capable of phosphorylating photon-activated rhodopsin and, thus, closely resembles rhodopsin kinase (Benovic et al. 1986). The sequences of the S-antigen and the arrestin of the β-adrenergic cascade (β-arrestin) are almost 60% identical (Lohse et al. 1990); at high concentrations the S-antigen can take over the function of the β-arrestin (Benovic et al. 1987). Based on these similarities it may be speculated that at least those "photoreceptor-specific" proteins which are expressed in the mammalian pineal organ in large quantities (e.g., the S-antigen) become involved in transmission of signals other than photic stimuli. This aspect is fascinating in regard to the evolution of transmembrane signaling processes and deserves further study.

Mammalian pinealocytes display an enriched glutamate immunoreactivity (McNulty et al. 1992; Redecker and Veh 1994, Fig. 8g). Moreover, microvesicles isolated from the bovine pineal gland have been shown to accumulate *L*-glutamate against a concentration gradient by means of a vesicular *L*-glutamate transporter (Moriyama and Yamamoto 1995a,b). These results suggest that microvesicles of mammalian pinealocytes accumulate glutamate, and that glutamate is secreted from pinealocytes upon exocytosis of microvesicles. Thus similarities between mammalian pinealocytes and true pineal photoreceptor cells appear to exist also in regard to the transmitter content. Interestingly, glutamate has been shown to affect the melatonin metabolism and the intracellular concentration of free calcium ions ($[Ca^{2+}]_i$) in mammalian pinealocytes (see below).

The close relationship between mammalian pinealocytes and neurons is stressed by the fact that the vast majority of mammalian pinealocytes are immunoreactive to synaptophysin, neurofilaments (Fig. 11c,d; Oksche et al. 1987; Vollrath and Schröder 1987; Redecker et al. 1990; Huang et al. 1992), synaptotagmin I, synaptobrevin II, syntaxin I (Redecker 1996), and rab 3 (Redecker 1995).

3.3
Central Innervation of the Pineal Organ

The central innervation of the pineal organ shows a striking variation among different classes of vertebrates but exists in all species studied so far. In principle, pinealofugal projections which originate from intrapineal neurons or in some cases also from pinealocytes are to be distinguished from pinealopetal projections that originate from various areas of the central nervous system and terminate within the pineal parenchyma.

3.3.1
Anamniotes and Reptiles

The pineal organ of anamniotes contains an intrapineal neuronal apparatus (Fig. 13) which receives synaptic inputs from true pineal photoreceptor cells and conveys these signals to the brain via a prominent pinealofugal projection (pineal tract and frontal organ nerve or parietal nerve). The intrapineal neurons have been investigated in many poikilothermic species by use of various techniques. Supravital staining with methylene blue has revealed bipolar, multipolar, and amacrinelike cells in the anuran pineal complex (Paul et al. 1971). Retrograde tracing experiments (Eldred and Nolte 1981; Ekström and Korf 1985; Blank et al. 1997) suggest that both multipolar and unipolar nerve cells contribute to the pineal tract (Fig. 13c). The histochemical demonstration of AChE has shown numerous neurons in the pineal of teleosts, amphibians, and reptiles, but most of them could not be classified because they did not display labeled processes (Wake 1973; Wake et al. 1974; Korf 1974, 1976; Korf et al. 1981; Vigh-Teichmann et al. 1982; Meissl and Ueck 1980; Fig. 13a). Thus the exact wiring diagram of the intrapineal neuronal network in poikilothermic vertebrates remains to be established.

Using immunocytochemical demonstration of neuropeptide Y (NPY) and neurofilament 200-kDa which are markers for horizontal and amacrine cells in the retina, Blank et al. (1997) did not find any immunoreactive neuronal perikaryon in the trout pineal organ and concluded that these cell types are missing from the pineal organ of the trout. These results suggest that the neuronal apparatus in the light-sensitive pineal organ is less complex than that of the retina. Results from tracing experiments are in line with this view: they show that intrapineal neurons which were retrogradely labeled via the pineal tract receive direct input from synaptic ribbon-containing terminals of true pineal photoreceptors (Fig. 13e). These findings argue in favor of a bineuronal chain in the pineal organ of the rainbow trout which comprises the photoreceptor cell as the first neuron and the ganglion cells forming the pineal tract as the second neuron. On the other hand, a few neuronal cells immunoreactive for GABA (Ekström et al. 1987b) and for substance P (Ekström and Korf 1986) have been found in the trout pineal organ, which may represent interneurons. The presence of a very limited number of interneurons may also be inferred from the demonstration of conventional synapses on intrapineal neurons.

The intrapineal neuronal network generates neuronal signals that can be recorded from the pineal tract or intrapineal ganglion cells and can be divided into two different response types: the chromatic and the achromatic (Figs. 14, 15; Dodt 1963, 1973; Morita 1966). The chromatic response is less frequent and shows chromatic antagonism between a short and a long wavelength mechanism. Light of short wavelengths in the visible and ultraviolet range elicits a long-lasting inhibition of the electrical activity of pineal ganglion cells, whereas light of longer wavelengths antagonizes the inhibitory response and causes excitation (Meissl and Dodt 1981; Solessio and Engbretson 1993; Meissl and Ekström 1993). The achromatic response consists of an inhibition of the

Fig. 13a–e. Intrapineal neurons and the pineal tract. **a** Acetylcholinesterase-positive neurons in the trout pineal organ. Sagittal section. ×250 **b** Neurofilament 200-kDa immunoreaction in the trout

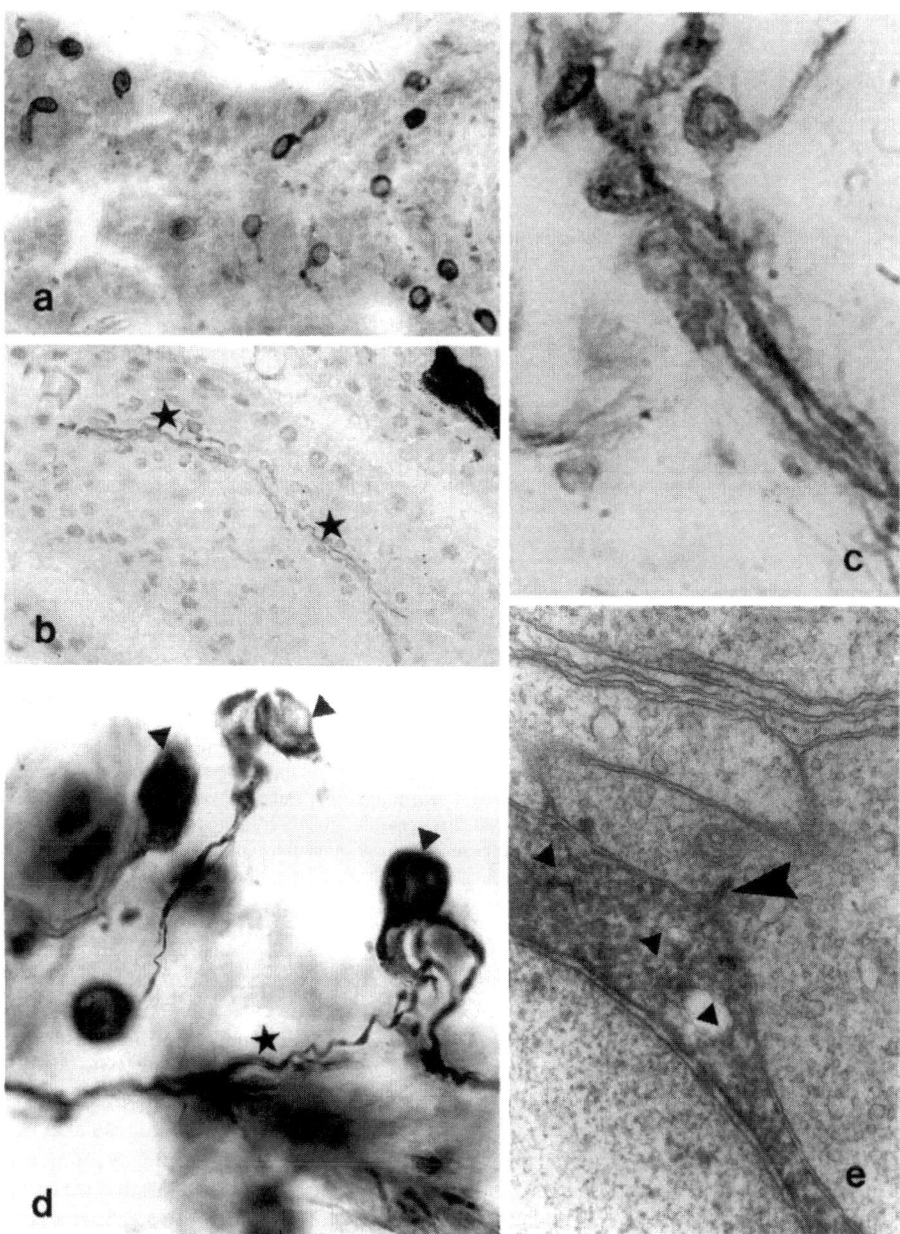

pineal tract (*stars*). Sagittal semithin section. ×120 (H. Blank, B. Müller, and H.-W. Korf, unpublished)
c Acetylcholinesterase-positive neurons in the pineal organ of the house sparrow. ×570 (Courtesy of
M. Ueck, Giessen; see Ueck and Kobayashi 1972) **d,e** Retrograde labeling of ganglion cells in the trout
pineal organ. The tracer DiI was applied to the proximal portion of the pineal stalk and photoconversion was performed as described by Blank et al. (1997). **d** Light micrograph of retrogradely labeled
ganglion cells (*triangles*) and axons of the pineal tract (*stars*). ×750 **e** Electron micrograph of a
retrogradely labeled ganglion cell (*triangles*) contacted by a process containing numerous clear
synaptic vesicles and a synaptic ribbon (*arrowhead*) ×30,000

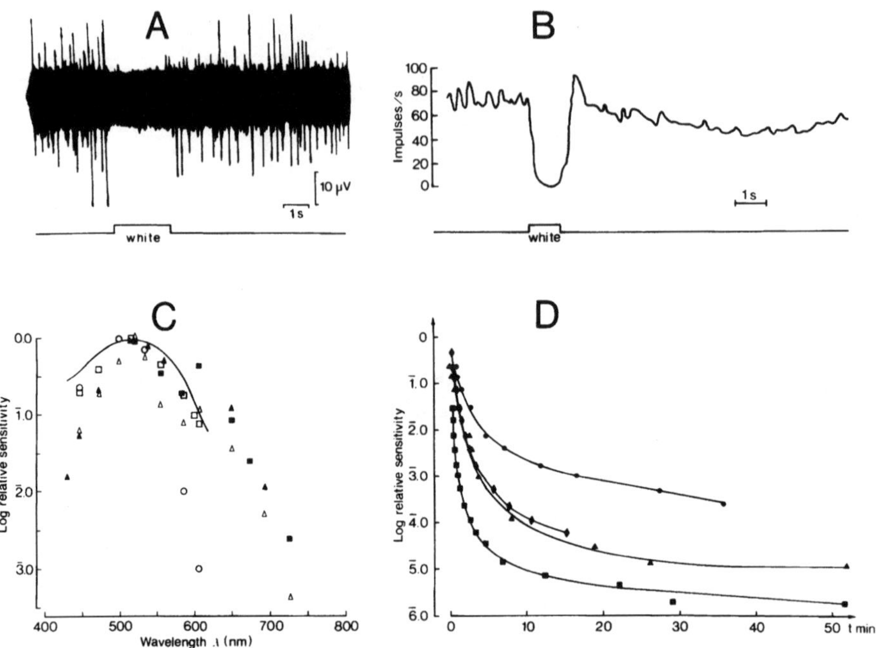

Fig. 14A–D. Achromatic responses of the frontal organ of the clawed toad, *Xenopus laevis*. **A** Action potentials recorded from the frontal organ nerve. Light stimulation is indicated by upward deflection of the lower beam. **B** Integrated spike frequency of mass spike potentials recorded from the frontal organ nerve. **C** Relative spectral sensitivity of the achromatic response determined by measurements of the relative quanta of light flashes of 1 s duration necessary for the just-perceptible inhibition of the impulse activity in the frontal organ nerve. *Continuous line*, Dartnall's nomogram absorption curve of visual pigment 520 nm. Different animals represented by different symbols. **D** Dark-adaptation curves obtained by a threshold criterion response. First measurement of each curve obtained a few s after cessation of a 20 min exposure to white light. (From Korf et al. 1981)

electrical activity upon stimulation with light of all wavelengths and an increase in electrical activity during darkness (Dodt 1963; Morita 1966).

The biological significance of these two different neuronal response has been discussed in some detail by Korf et al. (1981). The achromatic response supposedly acts as a dosimeter for solar radiation since, as shown for the frog, (a) constant illumination of the pineal organ reveals a linear relation between spike rate and the logarithm of luminance (Morita and Dodt 1965), and (b) the frequency of nerve impulses conducted to the brain via the pineal tract is related to the ambient light level. The chromatic response permits a color discrimination between long and short wavelengths of the visible spectrum. Meissl and Donley (1980) showed that the output of the frog frontal organ depends on a balance between opposing inhibitory and excitatory processes. This balance provides a very sensitive mechanism for the measurement of variation in the spectral distribution of the visible light. The natural photoperiod shows cyclic variations both in light intensity and spectral composition. The change in spectral distribution is most prominent during twilight. A decrease in a certain part of the visible spectrum is capable of shifting the chromatic response to another state

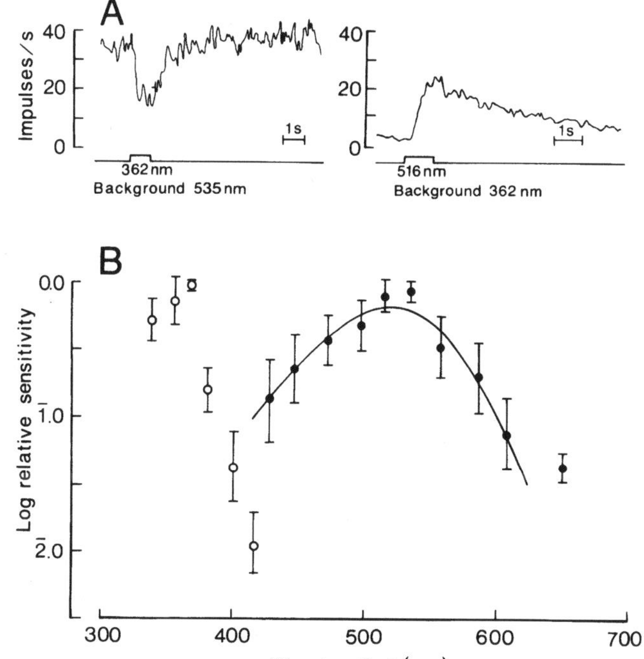

Fig. 15A,B. Chromatic response of the frontal organ of the clawed toad, *Xenopus laevis*. **A** Impulse rate time histogram for the inhibitory and excitatory component of the chromatic response. Both components were measured against an antagonistic background illumination. **B** Relative spectral sensitivity of the chromatic response: inhibitory action spectrum (*open circles*); excitatory spectrum (*dots*). The light threshold of the excitatory component is 2.2 log units lower than the inhibitory threshold. *Continuous line*, Dartnall's nomogram of the visual pigment 520. (From Korf et al. 1981)

of activation. Thus the chromatic response of the pineal may provide an important switch during twilight and, in addition to the achromatic system, act as a refined mechanism synchronizing physiological and behavioral activities.

In principle, the chromatic response is a characteristic feature of the extracranial component of the pineal complex, i.e., the frontal organ of amphibians and the parietal eye of lizards, which due to their superficial location are also capable of perceiving ultraviolet light. The achromatic response is the dominant neuronal signal of the intracranial pineal organ proper. The idea of a cooperative action of the chromatic and achromatic response in amphibians and lizards suggests that the extracranial and the intracranial components of the pineal complex act as a functional unit. This interpretation is supported by results from tracing studies showing that the neuronal projections from the frontal organ or parietal eye are identical with those from the pineal organ proper (Paul et al. 1971; Eldred et al. 1980; Korf and Wagner 1981).

The neuronal pathways of the pineal complex (frontal organ nerve; parietal nerve; pineal tract) may in total comprise as many as 3000 nerve fibers, most of which are unmyelinated (Korf 1974; McNulty 1984; see Vollrath 1981; Collin and Oksche 1981; Ekström and Meissl 1997). These nerve fibers constitute the pinealofugal (afferent) innervation that connects the anamniote pineal complex with the brain (see Korf and Oksche 1986). As shown for several species of teleosts, anurans, and lacertilians, the target areas of the pineal projections to the brain follow a basic pattern (*Bauplan*): they terminate within the reticular formation of the brainstem (central tegmental gray), pretectal area, habenular nuclei, several thalamic and hypothalamic nuclei, and the preoptic region. Notably, some of these areas also receive fiber inputs from the retina of the lateral eyes (Fig. 16; Korf and Wagner 1981; Ekström 1984). Such a dualistic innervation may guarantee the high precision in perception and transmission of one of the most important environmental zeitgeber, the photoperiod.

Interestingly, not only intrapineal second-order neurons but also certain pineal photoreceptor cells bear long axonal processes which leave the pineal organ via the pineal tract. Such long range projections of pinealocytes were first detected with the use an antiserum against the S-antigen in mammals (Korf et al. 1986b) and were subsequently found in teleosts (Ekström et al. 1987a). These initial observations were confirmed by use of tracing experiments and electron microscopy (Ekström 1987; Samejima et al. 1989). It remains to be established whether this type of pineal photoreceptor cell transmits its signals via graded potentials or generates and conducts action potentials.

The problem of whether the pineal complex of poikilothermic vertebrates is innervated by nerve fibers that originate from other brain regions is still a matter of discussion. Correlations of the number of AChE-positive nerve cells in the trout pineal organ with that of nerve fibers counted in the pineal tract of the same animals at the level of the subcommissural organ have revealed a surplus of approximately 300 fibers, suggesting the existence of pinealopetal axons in the pineal tract (Korf 1974). FMRFamide, gonadotropin releasing hormone, and NPY immunoreactive nerve fibers have been shown to innervate the teleost pineal (Ekström et al. 1988; Subhedar et al. 1996). The FMRFamide immunoreactive axons obviously originate from the nucleus of the terminal nerve which also innervates the retina (see Ekström and Meissl 1997). In a lizard the parietal eye has been shown to receive an innervation from the paraventricular nucleus (Korf and Wagner 1981); this projection seems to exist also in birds and mammals (see below). The functional significance of the pinealopetal nerve fibers remains enigmatic since all substances identified in such projections have very little if any effect on both the neuronal and the neuroendocrine activity of the pineal organ of poikilothermic vertebrates.

3.3.2
Birds

The basal processes of modified photoreceptors do not form synaptic contacts with intrapineal neurons but terminate freely at the basal lamina of the pineal parenchyma. Accordingly, the number of nerve cells in the pineal organ appears reduced in adult individuals of most avian species investigated. An apparent exception are passerine birds, whose pineal organ comprises several AChE-positive nerve cells and a well-developed pineal tract (Fig. 13c; Ueck and Kobayashi 1972; Korf et al. 1982). The functional role of this neuronal apparatus remains enigmatic because the sparrow pineal organ exerts its role within the photoneuroendocrine systems apparently via neuroendocrine mechanisms solely (Zimmerman and Menaker 1979).

Ontogenetic studies on the intrapineal neuronal apparatus in chicken and quail have revealed that the central innervation of the pineal (i.e., AChE-positive or neuron-specific enolase immunoreactive nerve cells) is reduced during posthatching development. The reduction is most prominent in the distal region of the pineal organ which is the first to be invaded by a dense sympathetic innervation (Sato and Wake 1983, 1984; Sato et al. 1990). The regression of the intrapineal neurons is paralleled by a

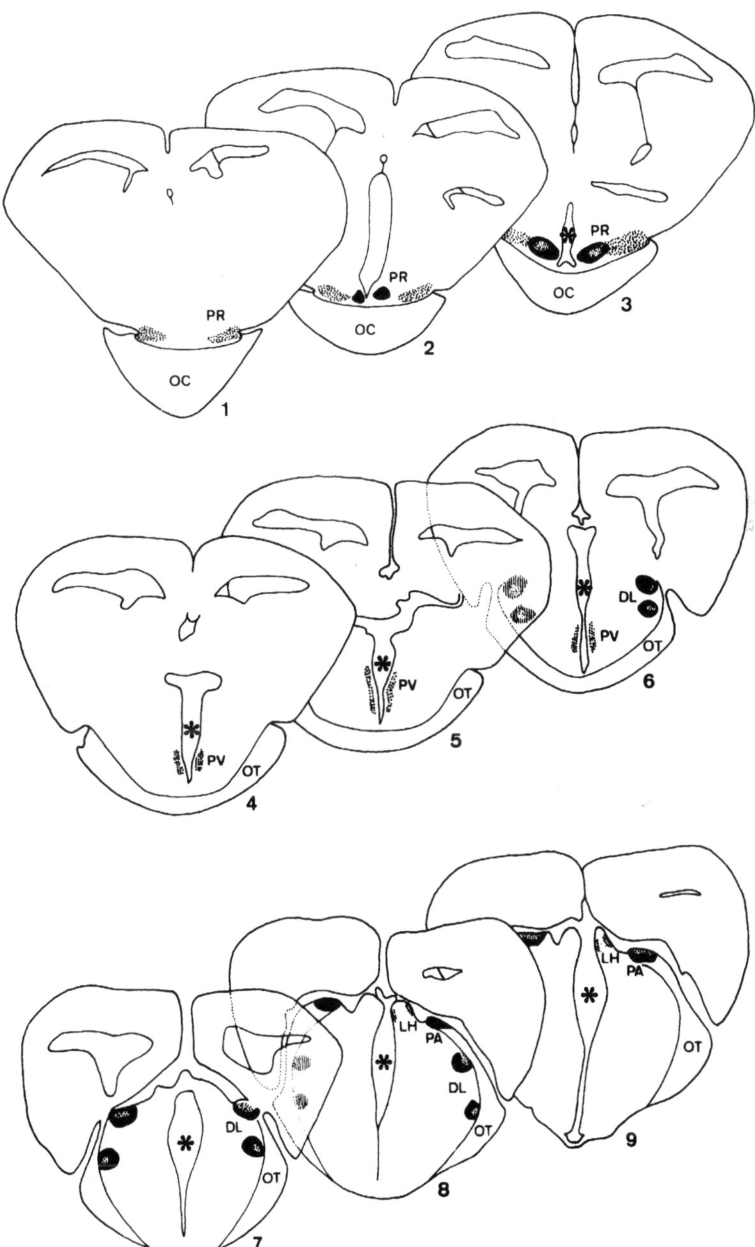

Fig. 16. Terminal areas of the parietal nerve (*dotted areas*) and both optic nerves (*filled areas*) in the brain of the lizard, *Lacerta sicula*, shown in a diagrammatic representation of frontal sections (*1*, rostralmost; *9*, caudalmost). *DL*, Dorsolateral thalamic nucleus; *LH*, left habenular nucleus; *OC*, optic chiasm; *OT*, optic tract; *PA*, pretectal area; *PR*, preoptic (tel- and diencephalic) region; *PV*, periventricular gray; *asterisk*, third ventricle. Note partial overlapping of parietal nerve and optic nerve projections and the asymmetric location of the parietal nerve in the left habenular nucleus. (From Korf and Wagner 1981)

regression of pinealocytes immunoreactive for photoreceptor-specific proteins (Korf 1994). Thus an inverse relationship seems to exist between the development and differentiation of the photoreceptive/neuronal apparatus and the sympathetic innervation of the organ.

Tracing studies have demonstrated that the pineal organ of the house sparrow receives a central pinealopetal innervation that originates from the habenular nuclei and the periventricular gray of the hypothalamus (Korf et al. 1982).

3.3.3
Mammals

The central innervation of the mammalian pineal organ appears regressed in comparison to that of anamniotes. In several mammalian species neurons have been observed as a part of the pineal parenchyma; but species differences are evident with regard to their number. They are extremely rare or even absent in rodents (see Korf and Møller 1984, for review and references). In contrast, accumulations of intrapineal nerve cells have been found by means of various techniques in adult individuals of other mammalian species (ferret: AChE activity and electron microscopy, Trueman and Herbert 1970; David and Herbert 1973; cat: GABA immunoreaction and electron microscopy, Vigh-Teichmann et al. 1991; rabbit: silver impregnation and electron microscopy, Romijn 1973, 1975; bovine: choline acetyl-transferase immunoreaction, Phansuwan-Pujito et al. 1991; rhesus monkey: electron microscopy, David et al. 1975; humans: enkephalin immunoreaction, Moore and Sibony 1988). The origin and function of these neurons remain enigmatic. Some investigators (Trueman and Herbert 1970; Korf and Møller 1984, 1985; Vigh-Teichmann et al. 1991) consider most of the cells as derivatives of the central nervous system which are engaged in communication between the pineal and the brain, and which closely resemble the nerve cells found within the sensory pineal organ of lower vertebrates. Other authors (Romijn 1973, 1975; Møller 1992) have argued that such nerve cells belong to the parasympathetic system (see below).

The central innervation of the mammalian pineal organ is most prominent in earlier stages of ontogenetic development: intrapineal nerve cells are more numerous in human fetuses than in adults. Moreover, a solid bundle of nerve fibers connecting the pineal organ with di- and mesencephalic areas has been observed in human, rabbit and sheep fetuses, but is no longer detectable in adult individuals (see Møller 1992, for review and references).

An additional type of direct connection between the pineal and the brain has been detected in hamster and mouse. This is established by long axonlike processes of S-antigen immunoreactive pinealocytes which enter the habenular nucleus and the pretectal area (Fig. 17; Korf et al. 1986b, 1990). These findings suggest that signals from the mammalian pineal organ are conveyed not only by means of neuroendocrine mechanisms (i.e., biosynthesis and release of melatonin into the general circulation) but also via neuronlike pathways. Whether the latter mode of signal transmission is under the control of the sympathetic nervous system as is the melatonin biosynthesis remains to be established. Certainly, pinealocyte processes within the habenular nucleus receive direct input from the central nervous system, because they form the postsynaptic part of conventional synapses, whose presynaptic terminal contains

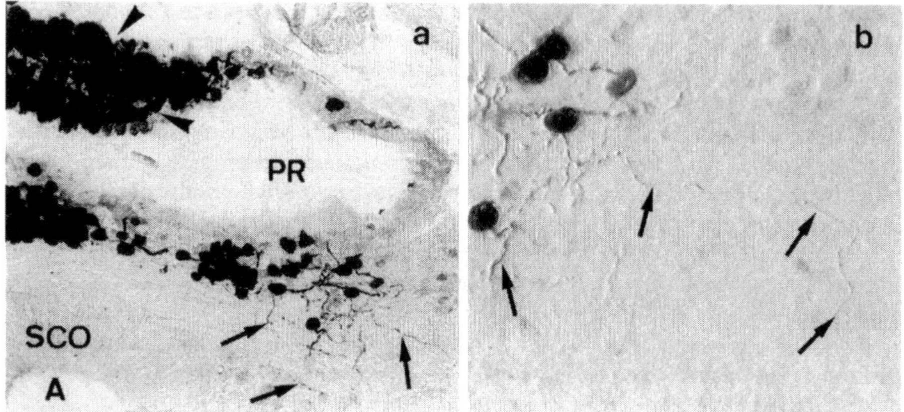

Fig. 17.a,b S-antigen immunoreactive processes of mouse pinealocytes which penetrate into the habenular region. Frontal sections through the deep pineal organ, habenular region and subcommissural organ. *PR*, Pineal recess; *SCO*, subcommissural organ; *A*, cerebral aqueduct; *arrows*, S-antigen immunoreactive processes. a ×180; b ×380 (From Korf et al. 1990)

numerous clear vesicles and belongs to the central "pinealopetal" innervation (Korf et al. 1990).

The existence of a central pinealopetal innervation in mammals has already been demonstrated by several early investigators (see Bargmann 1943; Korf and Møller 1984, 1985; for review and references). This type of innervation received renewed attention after its existence has been confirmed by means of anterograde and retrograde tracing techniques as well as lesion experiments (Korf and Møller 1984, 1985; Møller 1992). Nerve fibers innervating the rodent pineal organ have been found to originate from the medial and lateral habenular nuclei, hypothalamic paraventricular nucleus, intergeniculate leaflet of the lateral geniculate body (Mikkelsen and Møller 1990; Mikkelsen et al. 1991), pretectal area, and nucleus of the posterior commissure. Additional nerve fibers may originate from serotonergic and adrenergic neurons in the central nervous system (Matsuura and Sano 1983; Zhang et al. 1991). Notably, the central innervation of the pineal organ is not as conspicuous as the sympathetic innervation and, in rodents, comprises less than 100 nerve fibers. There is, however, some evidence that nonrodents (sheep, cow, dog, cat, monkey, humans) are endowed with a more prominent central pinealopetal innervation (Møller 1992). The density of the central innervation may thus be related to the morphological type of the pineal organ (Møller 1992).

3.4
Sympathetic (Noradrenergic) Innervation of the Pineal

The sympathetic innervation shows a progressive development in the course of evolution and is of utmost importance for the regulation of the melatonin biosynthesis and other pineal functions in mammals. Norepinephrine (NE), the primary sympathetic neurotransmitter, stimulates melatonin biosynthesis in all mammalian species.

Notably, NE has an inhibitory effect on the melatonin biosynthesis in all avian species investigated to date. This conversion occurs at the level of the adrenoreceptors (see below).

The progressive development of the sympathetic innervation appears to coincide with the reduction of true pineal photoreceptors and the central neuronal apparatus. Such a coincidence suggests that the sympathetic innervation also influences the phenotype of the pineal organ. This view gains support from results obtained with various species of birds, showing that during ontogeny, the central innervation of the pineal (represented by AChE-positive nerve cells) is reduced in the most distal part of the pineal organ which receives a dense sympathetic innervation (Sato and Wake 1983; Sato et al. 1990; see p. 35, 49ff).

Another interesting effect of NE is related to the expression of immunoreactive rod-opsin in pinealocytes of albino rats (Araki 1992; Araki and Tokunaga 1990; Araki et al. 1988). In contrast to other mammals, albino rodents (rats and mice) display no or extremely few rod-opsin immunoreactive pinealocytes when the pineal is investigated in situ (Korf et al. 1985a; Korf and Ekström 1987). Araki et al. (1988) noted a drastic increase in number of rod-opsin immunoreactive cells after pineal glands of newborn albino rats were transferred to cell culture conditions; the authors assumed that the elevated expression of immunoreactive rod-opsin was due to a denervation effect, and that NE suppresses the differentiation of cells with photoreceptor properties. This view gained support from experiments involving NE treatment of cultured pinealocytes isolated from newborn albino rats (Araki and Tokunaga 1990). The addition of NE resulted in a nearly total reduction of rod-opsin immunoreactive cells. Moreover, the extension of neuritelike processes (see Korf et al. 1986b) was suppressed upon addition of NE to the culture medium. Other biogenic amines such as dopamine or serotonin show no such inhibitory effects.

Subsequently the influence of NE on development of photoreceptor and neuronal characteristics (for definition, see Oksche et al. 1987) was studied in greater detail (Araki 1992). There seem to be critical stages (between postnatal days 1 and 4) in which NE treatment can influence the expression of rod-opsin immunoreactivity. Cultures taken from the pineal of older rats do not show this inhibitory phenomenon. The suppressive effects of NE on the expression of rod-opsin immunoreactivity are mimicked by adding a depolarizing agent, 5–10 mM KCl. In conclusion, these results suggest that NE, in addition to influencing the metabolic activity of the pineal organ of adult mammals, is involved in regulating the development and final phenotype of the organ. This aspect is particularly intriguing since the pineal complex of lower vertebrates, which is endowed with a well-developed photoreceptor and neuronal apparatus (see Vollrath 1981; Korf and Oksche 1986; Korf and Ekström 1987), is less densely innervated by sympathetic noradrenergic nerve fibers (see below). On the other hand, a considerable number of rod-opsin immunoreactive pinealocytes are present in adult mammals other than albino rodents, despite the fact that their pineal organ receives a dense sympathetic innervation.

3.4.1
Anamniotes

In the pineal organ of poikilothermic vertebrates, photoreception and phototransduction appear to be the prevailing regulatory mechanism. Accordingly, a noradrenergic (sympathetic) innervation of the pineal organ seems to be absent or at least much less developed in anamniotes than in mammals and birds. With the Falck-Hillarp technique green fluorescent (probably noradrenergic) nerve fibers have been found in the meninges surrounding the pineal organ of the pike, *Esox lucius*, but these fibers do not enter the pineal parenchyma (Owman and Rüdeberg 1970); a similar arrangement has been observed in ranid frogs (Owman et al. 1970). More recently Blank et al. (1997) have demonstrated NPY immunoreactive nerve fibers in the capsule and the perivascular space of the trout pineal organ. This finding suggests that roots of the autonomic pinealopetal innervation can be traced back to fishes since NPY and NE are known to be colocalized in sympathetic neurons and nerve fibers (Lundberg et al. 1982; see p. 41). This holds also true for the sympathetic nerve fibers innervating the mammalian pineal gland (Schröder 1986; Schröder and Vollrath 1986; Shiotani et al. 1986; Reuss and Moore 1989; Cozzi et al. 1992; Møller et al. 1990, 1994).

3.4.2
Reptiles

Green fluorescent nerve fibers are more numerous in the reptilian pineal organ than in that of anamniotes. They are seen in the parenchyma and the perivascular space of the pineal organ (Quay et al. 1968; Wartenberg and Baumgarten 1969). Several lines of evidence suggest that such fibers are of sympathetic origin and contain both NE and serotonin. In the ophidian pineal organ numerous green fluorescent (noradrenergic) nerve fibers run in the perivascular space. A few of them enter the pineal parenchyma. In *Natrix* these fibers have been shown to be largely derived from a dorsal and posterior meningeal nerve bundle reminiscent of the conarian nerve in mammals. The functional significance of the sympathetic innervation of the reptilian pineal organ remains to be established.

3.4.3
Birds

Fluorescence microscopy and microspectrofluorometry have demonstrated noradrenergic nerve fibers in several species of birds (Fig. 18a; see Ueck 1979; Vollrath 1981; for review and references). The number and intrapineal location of these nerve fibers vary with the species. In passerine birds, whose pineal comprises a relatively well-developed photosensory and neuronal apparatus (Ueck and Kobayashi 1972; Menaker and Oksche 1974), green fluorescent fibers are located preferentially in the perivascular spaces, whereas in the duck and pigeon they penetrate into the pineal parenchyma (Ueck 1973). The noradrenergic nerve fibers innervating the avian pineal obviously originate from the SCG: they disappear after bilateral removal of the SCG (Hedlund and Nalbandov 1969). The noradrenergic nerve fibers of the avian pineal contain

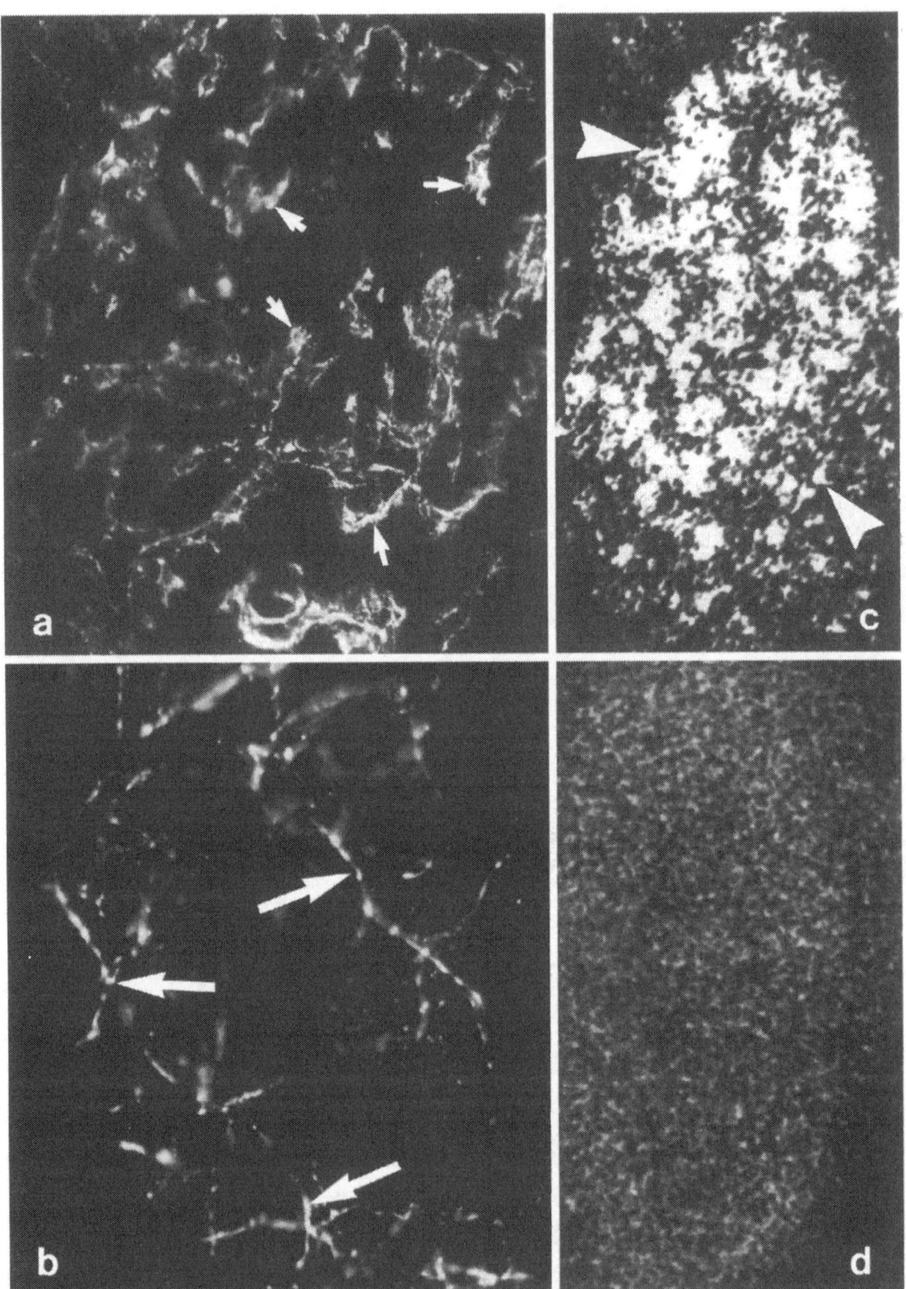

Fig. 18a–d. Autonomic innervation of the pineal organ. a Green fluorescent (noradrenergic) nerve fibers (*arrows*) in the pineal organ of the tree sparrow, *Passer montanus*, as visualized by the Falck-Hillarp technique. ×95 (Courtesy of M. Ueck, Giessen; see Ueck 1973) b Immunocytochemical demonstration of the serotonin-transporter in sympathetic nerve fibers (*arrows*) of the rat pineal organ. ×600 c Demonstration of mRNA for the serotonin transporter in nerve cells (*arrowheads*) of the rat superior cervical ganglion by means of in situ hybridization and an antisense cDNA probe (see

characteristic dense-core granules, and in regard to their ultrastructure they are very similar to those observed in the mammalian pineal organ (Ueck 1970).

In vitro studies suggest that the sympathetic input is not mandatory for the maintenance of the rhythm in the melatonin biosynthesis in the avian pineal organ. Although the sympathetic innervation of the pineal is well developed in birds, its functional significance is not quite clear, and there may be considerable differences among various avian species. Notably, the action and turnover of NE are strikingly different in birds and mammals. In birds NE inhibits the melatonin biosynthesis, and its turnover is high during the day and low at night and is thus out of phase with the melatonin rhythm by 180° (Cassone et al. 1986, see below).

3.4.4
Mammals

Of all vertebrate classes mammals possess the most prominent sympathetic innervation of the pineal organ. This type of innervation was first described by Cajal (1904) in the mouse pineal organ. Cajal showed nerve fibers which enter the surface of the mouse pineal gland together with blood vessels and extend into the parenchyma from a perivascular location. A similar innervation of the pineal organ was found in several other mammalian species, for example, rabbit (Pines 1927), cat (Antonow 1925), dog (Hartmann 1957), and cattle (Antonow 1925). In 1922 Kolmer and Löwy observed a nerve bundle located between the vena cerebri magna of Galen and the tip of the pineal gland in dog, goat, and humans; they called this bundle the "nervus conarius." Subsequently this nerve was shown in the macaque (Le Gros Clark 1940). In 1960 Kappers demonstrated the conarian nerve in the rat and showed it to be a bilateral structure and to degenerate after bilateral removal of the SCG, thus verifying its sympathetic nature and origin from the SCG. These findings led to the concept that in mammals the sympathetic nerve fibers from the SCG form the only innervation of the pineal gland (Kappers 1960, 1965). Although this view has been proven too simplistic in the meantime (see Korf and Møller 1984, 1985; Møller 1992), it provided a most valuable and stimulating morphological basis for a series of elegant biochemical and physiological experiments which elucidated the interactions between the postganglionic sympathetic nerve fibers and the neuroendocrine activity of the mammalian pineal gland.

Already the classical neurohistological studies by Cajal (1904) and Kappers (1960, 1965) clearly demonstrated that the sympathetic innervation of the pineal organ originates from the SCG. Retrograde tracing experiments (Bowers et al. 1984; Reuss and Schröder 1988) confirmed these observations. After injection of horseradish peroxidase into the rat pineal gland and lesion of one conarian nerve 250 neurons were labeled in the SCG ipsilateral to the intact conarian nerve. The labeled neurons were distributed all over the ganglion; however, they appeared to be more concentrated in

◄——————————————————————

Pfeffer et al. 1997). The location of the labeled neurons corresponds to that of nerve cells projecting to the pineal organ (see Bowers et al. 1984). **d** No label is found in the rat superior cervical ganglion after application of a sense-probe for the serotonin-transporter. **c,d** ×55 (M. Pfeffer, J.H. Stehle, P. Schloss, H. Betz, and H.-W. Korf, unpublished)

the rostral quadrants (Bowers et al. 1984). The sympathetic pinealopetal nerve fibers, which finally reach the gland via the conarian nerves, leave the SCG through the internal carotid nerve. The tracing study conforms to previous investigations which revealed that (a) bilateral stimulation of the internal carotid nerve leads to an increase in pineal NAT activity comparable to that observed after stimulation of the preganglionic sympathetic trunk (Bowers and Zigmond 1982), and (b) bilateral lesion of the internal carotid nerve abolishes the nighttime rise in the activity of this enzyme (Zigmond et al. 1981).

It is generally accepted that both SCG innervate the pineal organ to the same extent (Bowers and Zigmond 1982). Moreover, nerve fibers from left and right SCG interact within the pineal gland since unilateral stimulation of preganglionic nerve fibers results in an increase in NAT activity which is less than half the increase observed after bilateral stimulation (see Korf 1996, for further discussion of this aspect). Using the fluorogold technique Reuss and Schröder (1988) have found that approximately one-third of the SCG cells projecting to the pineal are in juxtaposition with "small intensely fluorescent" cells, which are considered as interneurons or as elements with a special paracrine capacity. The functional relationship of this close apposition, however, remains enigmatic to date.

The overall distribution of aminergic nerve fibers in the pineal organ has been extensively studied by use of the formaldehyde-induced histofluorescence method developed for the detection of catecholamines and indoleamines by Falck et al. (1962). Since nearly all fluorescent fibers present in the mammalian pineal disappear after superior cervical ganglionectomy, they are considered as sympathetic axons. As shown for various species of mammals, these nerve fibers mostly exhibit a green fluorescence indicating that they contain NE as neurotransmitter (see Vollrath 1981 for review and a complete list of references). The fluorescence microscopic findings have been confirmed by means of microspectrofluorometric analyses (Møller et al. 1979). In some mammalian species (e.g., rat, gerbil, hamster) the intrapineal sympathetic nerve fibers also contain serotonin (Fig. 12c; Owman 1964; Nielsen and Møller 1978; Sheridan and Sladek 1975). This is probably due to an uptake of serotonin released from pinealocytes (Zweig and Axelrod 1969), which may be regulated via a recently cloned serotonin-transporter. This molecule has been demonstrated immunocytochemically in fiberlike profiles of the rat pineal gland which according to their course and arrangement resemble the sympathetic nerve fibers (Fig. 18b). mRNA encoding for the serotonin transporter was not detected in the rat pineal organ, but it was present in high abundance in several perikarya of the SCG (Fig. 18c). The location of the labeled ganglion cells was strikingly similar to that of ganglion cells which were retrogradely labeled after tracer injection into the rat pineal organ (Bowers et al. 1984).

The sympathetic nerve fibers enter the pineal either via the conarian nerves or along pial blood vessels. Within the pineal gland they mainly follow the vascular system. However, several fibers leave the perivascular space and penetrate into the parenchyma where they are found between the pinealocytes. Such a rich parenchymal innervation has been observed in the rat (Owman 1965), cat, and gerbil (Nielsen and Møller 1978), whereas in the guinea pig the nerve fibers are found predominantly in the perivascular space (Owman 1965). In certain rodents (e.g., the rat) the sympathetic nerve fibers extend from the pineal parenchyma into the medial habenular nucleus (Björklund et al. 1972). This peculiar innervation reinforces the concept that the medial habenular nucleus represents a transitional area which contains pinealocytes and conventional neurons and connects the pineal organ to the diencephalic roof (Korf et al. 1990, see p. 32).

Catecholaminergic nerve fibers in the pineal organ have also been visualized by means of immunocytochemical demonstration of tyrosine hydroxylase (Zhang et al. 1991) and dopamine-β-hydroxylase (Fig. 19a; see also Schröder and Vollrath 1985). In the rat nearly all tyrosine hydroxylase immunoreactive fibers disappear from the superficial pineal organ after bilateral superior cervical ganglionectomy, and only a few remain which may originate from central noradrenergic neurons (see above). Very similar immunocytochemical results have been obtained with the sheep pineal organ (Cozzi et al. 1992).

Electron microscopy has shown that the sympathetic nerve fibers form terminals in which the majority of transmitter vesicles are 40–60 nm in diameter and include a small, often eccentric electron dense core (Arstila 1967). In addition, a few larger granular vesicles (approximately 100 nm in diameter) may be present (Fig. 20). Several investigations have indicated that the small dense core granules contain NE. The dense core disappears after treatment with reserpine known to deplete the NE pool of the nerve fibers (Pellegrino de Iraldi et al. 1965; Arstila 1967; Wood 1973). The electron density of the granule increases after treatment of the animals with NE precursors. The number of dense cores in the transmitter vesicles has been shown to decrease after electrical stimulation of the SCG as a result of the release of catecholamines confined in the vesicle (Jaim-Etcheverry and Zieher 1980). Electron microscopic histochemistry has also provided evidence that monoamines are located within the dense core of the small granular vesicles (Jaim-Etcheverry and Zieher 1971). The small (and also certain large granular) vesicles of the sympathetic nerve fibers are capable of uptake of exogenously administered monoamines (Jaim-Etcheverry and Zieher 1983).

Electron microscopic studies show that sympathetic nerve terminals are located in both perivascular and intraparenchymal positions, thus confirming the fluorescence microscopic results on the location of sympathetic nerve fibers. Most sympathetic nerve fibers terminate freely, and only a few establish specialized, synapselike contacts with pinealocytes or other cellular elements in the pineal (Kappers 1969; Huang and Lin 1984; Ling et al. 1990). Investigations with the use of antibodies against synaptophysin, a membrane protein of clear synaptic vesicles in nerve fibers and neuroendocrine cells, have revealed that mammalian pinealocytes extend processes into the perivascular space where they form several terminal swellings containing numerous clear synaptophysin-immunoreactive vesicles. These terminal swellings are in very close spatial relationship with nerve fibers which on the basis of ultrastructural criteria can be classified as sympathetic nerve fibers (Redecker 1993). Consequently the regulation of the metabolic activity by sympathetic nerve fibers may occur at the level of pinealocyte processes.

Although point-to-point contacts between pinealocytes and noradrenergic sympathetic nerve fibers appear to be rare, they may be of importance for the regulation of pineal functions. Impulses transmitted from the sympathetic nerve fiber to selected pinealocytes through such point-to-point contacts may then be conveyed to other pinealocytes via gap junctions shown to exist between mammalian pinealocytes (Taugner et al. 1981; Huang and Taugner 1984) and to establish a functional pathway for electric and metabolic coupling between pinealocytes (Sáez et al. 1991). Interestingly, coupling between pinealocytes is influenced by NE, as shown with cultured rat pinealocytes (Sáez et al. 1991). Treatment of the cultured pinealocytes with NE in-

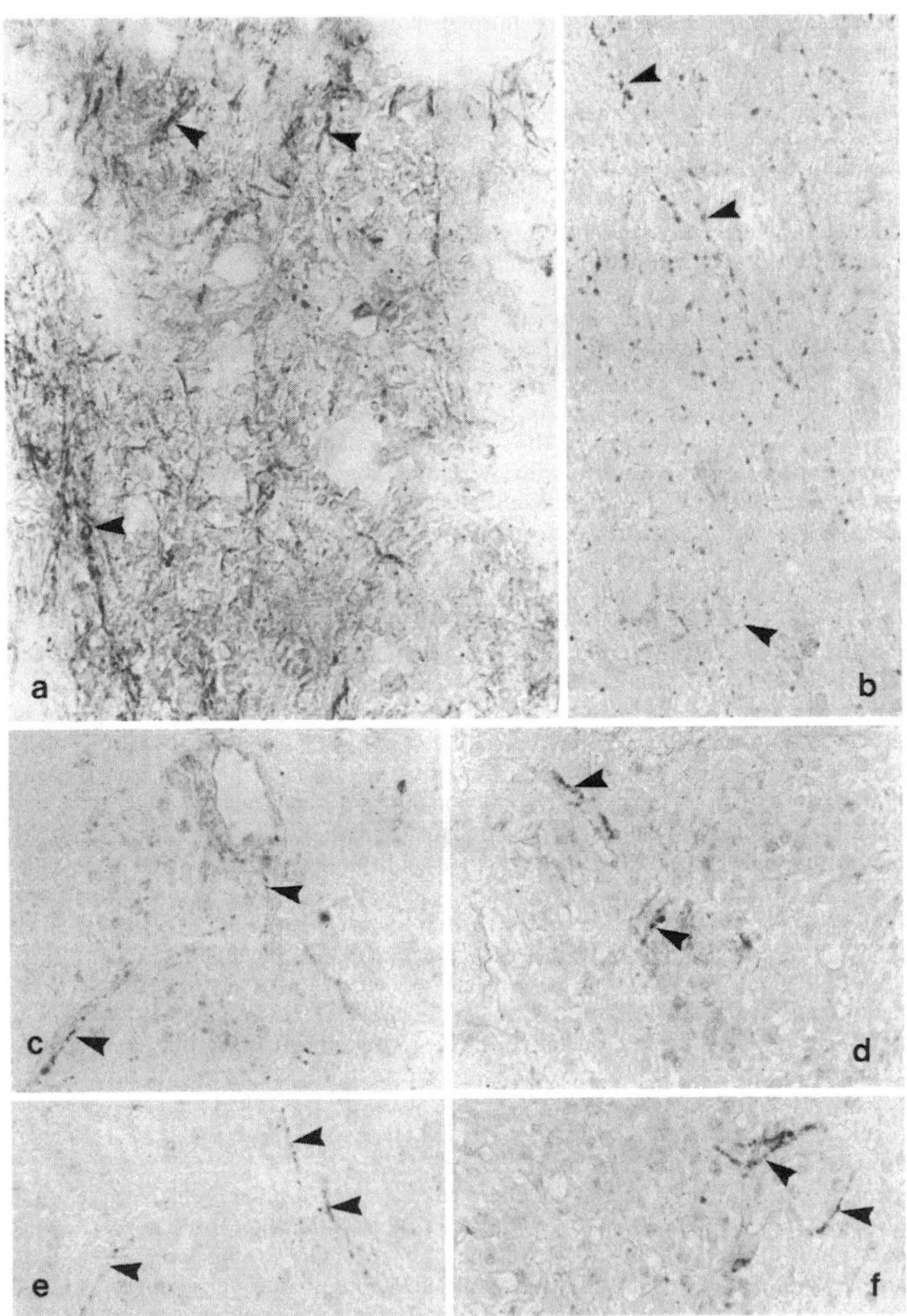

Fig. 19a–f. Immunocytochemical demonstration of catecholaminergic and peptidergic nerve fibers (*arrowheads*) in the mammalian pineal organ. **a** Dopamine-β-hydroxylase (*DBH*) immunoreactivity labeling noradrenergic nerve fibers in the pineal organ of an adult human. Antiserum against DBH was a generous gift from E. Rodriguez, Valdivia, Chile. ×300 **b–f** Sheep pineal organ. **b** Neuropeptide-Y immunoreactive nerve fibers. ×300 **c,e** Vasoactive intestinal peptide immunoreactive nerve fibers. ×300 **d,f** Substance P immunoreactive nerve fibers. ×300

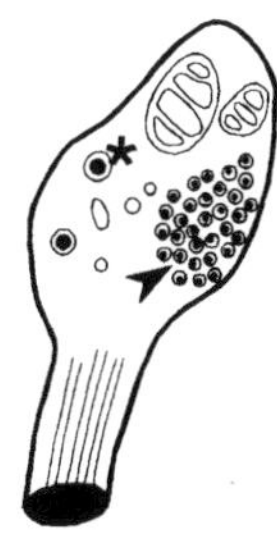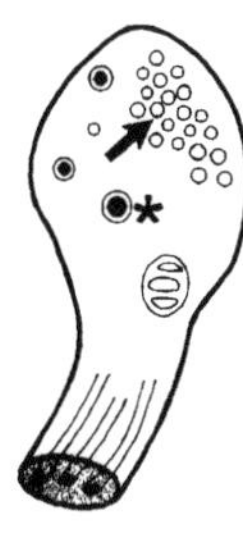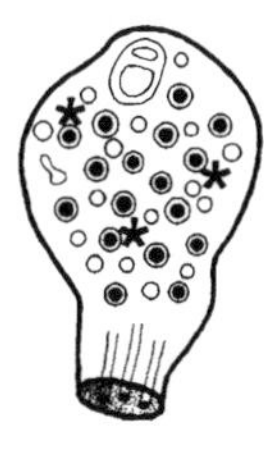

Fig. 20. Schematic drawing of the three different types of nerve terminals observed in the mammalian pineal organ by transmission electron microscopy. The sympathetic nerve terminals contain numerous vesicles (40–60 nm) often confining an eccentric dense core (*arrowhead*) and some larger dense-core vesicles (*star*). The cholinergic nerve terminal contains numerous clear synaptic vesicles (*arrow*) and some larger dense-core vesicles (*star*). The peptidergic terminal contains numerous large dense-core vesicles (*stars*).

creases the incidence of dye coupling and the level of immunodetectable connexin (Sáez et al. 1991). The NE effect is paralleled by an increase in immunodetectable connexin and prevented by blockade of mRNA or protein synthesis. Thus the induction of coupling appears to be mediated by synthesis of gap junction protein and not through a gating process. It remains to be determined whether the increase in dye coupling is causally related to the induction of melatonin biosynthesis. Interestingly, the maximal effects of NE on dye coupling in rat pinealocytes has been observed 6 h after NE treatment. A similar time lag has been found in the rat between the release of NE and the maximal induction of NAT.

From other organs and systems it is known that sympathetic neurons and nerve fibers contain NPY in addition to the classical neurontransmitter NE (Lundberg et al. 1982; Ekblad et al. 1984). Immunocytochemistry has also shown NPY in nerve fibers and terminals in the pineal organ of various mammalian species (rat: Schon et al. 1985; Zhang et al. 1991; hamster: Schröder 1986; gerbil: Shiotani et al. 1986; guinea pig: Schröder and Vollrath 1986; mink: Møller et al. 1990; cat: Møller et al. 1994; sheep: Williams et al. 1989; Cozzi et al. 1992; Fig. 19b). Most of these NPY immunoreactive nerve fibers appear to belong to the sympathetic innervation since they degenerated after bilateral superior cervical ganglionectomy (rat: Zhang et al. 1991; sheep: Cozzi et al. 1992; cat: Møller et al. 1994). Additional evidence for the sympathetic origin of NPY immunoreactive nerve fibers derives from studies combining tracing techniques and immunocytochemistry (Shiotani et al. 1986; Reuss and Moore 1989). Some NPY immunoreactive nerve fibers, however, persist in the pineal organ of rats, cats, and sheep after ablation of both SCG (Zhang et al. 1991; Cozzi et al. 1992; Møller et al. 1994). These elements most probably originate from diencephalic areas and represent a part of the central pinealopetal innervation (see Korf and Møller 1984, 1985). Alternately, it has been suggested that they originate from parasympathetic or even sensory ganglia (see below).

3.5 Parasympathetic Innervation of the Pineal Organ

Very little is known about the parasympathetic innervation of the pineal organ, and all of the limited data have been obtained with mammals. As discussed above, some researchers consider the nerve cells found in the pineal organ of some mammalian species as parasympathetic neurons (Romijn 1973, 1975; Møller 1992). To clarify further the function and origin of these nerve cells it is important to determine whether they are influenced by preganglionic axons originating from central visceromotor nuclei. Romijn (1973, 1975) suggested that the intrapineal nerve cells in the rabbit receive preganglionic input from the superior salivatory nucleus via the conarian nerve. Kenny (1961) showed in the monkey that lesions of the greater petrosal nerve are followed by degeneration of (a) nerve fibers running in the conarian nerve and terminating in the pineal parenchyma and (b) of intrapineal nerve cells. According to Kenny (1961) and Romijn (1975), the conarian nerve comprises not only postganglionic sympathetic nerve fibers but also pre- and postganglionic parasympathetic nerve fibers. For the sake of clarity it seems desirable to re-examine this type of innervation with the use of modern neurobiological techniques, for example, tracing methods.

After the injection of fluorescent tracers into the gerbil pineal organ Shiotani et al. (1986) observed labeled perikarya in the pterygopalatine ganglion and concluded that some of the peptidergic nerve fibers innervating the pineal organ (see below) represent postganglionic parasympathetic nerve fibers. The precise course of these fibers, however, remains to be elucidated. In the same study Shiotani et al. (1986) also observed labeled perikarya in the trigeminal ganglia. Fibers originating from this source, however, are more likely to provide the sensory innervation of the dura than to innervate the pineal parenchyma.

In some species, for example, humans, collections of nerve cells have been shown to accumulate in ganglia located in close topographical relation with the pineal organ but outside the pineal parenchyma. One ganglion (termed Marburg's ganglion) may be found at the most distal portion (apex) of the gland above the great cerebral vein of Galen; another ganglion (Pastori's ganglion) may occur beneath the great cerebral vein (Bargmann 1943; Vollrath 1981; Møller 1992). These ganglia, however, appear as inconsistent structures; some authors have failed to detect them (see Bargmann 1943). Nothing is known about the neurotransmitters employed by such neurons; also their function remains enigmatic. Møllgard and Møller (1973) believe that they belong to the parasympathetic system.

According to the classical concept on parasympathetic neurons and nerve fibers, they employ acetylcholine (ACh) as the primary neurotransmitter. The presence of cholinergic nerve fibers in the mammalian pineal organ has been repeatedly investigated. Phansuwan-Pujito et al. (1991) demonstrated immunoreactive choline acetyltransferase, the rate-limiting enzyme in ACh synthesis, in axons innervating the bovine pineal organ. The origin of these cholinergic nerve fibers has not yet been confirmed experimentally; it is suggested that they represent parasympathetic nerve fibers (Phansuwan-Pujito et al. 1991). Alternately, such elements may be of central origin. Further evidence for a cholinergic innervation of the mammalian pineal organ has been obtained by studies using antibodies against the vesicular ACh transporter (Weihe et al. 1996). Recent data from calcium imaging studies (Schomerus et al. 1995; Korf et al. 1996) and patch-clamp recordings (Letz et al. 1997) support the notion that these cholinergic nerve fibers are relevant for regulation of pineal functions in mammals.

3.6 Peptidergic Innervation of the Pineal Organ

Information on peptidergic nerve fibers in the pineal organ of poikilothermic vertebrates and birds is scarce. FMRFamide, gonadotropin releasing hormone, and NPY
immunoreactive nerve fibers have been shown to innervate the fish pineal organ
(Ekström et al. 1988; Subhedar et al. 1996; Blank et al. 1997; see also p. 30), but their
functional role for regulation of pineal functions in fish remains elusive. With regard
to birds Pratt and Takahashi (1989) have reported that sparse vasoactive intestinal
peptide (VIP) immunoreactive nerve fibers are located within the connective tissue
capsule surrounding the pineal follicles in the chick. This conforms to biochemical
results showing that VIP regulates cAMP in chicken pinealocytes (see below). The
origin of these VIP immunoreactive nerve fibers remains to be determined.

The peptidergic innervation of the mammalian pineal organ is diversified with
regard to the neuropeptide content and origin. Peptidergic nerve fibers in the pineal
may originate from different sources: (a) from the SCG of the sympathetic trunk, (b)
from various diencephalic and mesencephalic nuclei, (c) from parasympathetic or
sensory ganglia, and (d) from intrapineal neurons or peptidergic pinealocytes.

As discussed above, NPY mainly appears to be colocalized with NE in sympathetic
nerve fibers. Apart from NPY, which may represent the most widespread neuropeptide
in intrapineal nerve fibers, several other neuropeptides have been located immunocytochemically in nerve fibers projecting to and terminating within the pineal organ
(Fig. 19c–f). These include the C-flanking peptide of NPY (cat: Møller et al. 1994; sheep:
Cozzi et al. 1992), substance P (hedgehog: Korf and Møller 1985; rat: Ronnekleiv and
Kelly 1984; Korf and Møller 1985; gerbil: Shiotani et al. 1986; bovine: Møller et al. 1993;
macaque: Ronnekleiv 1988), somatostatin (bovine: Møller et al. 1992), VIP (rat: Mikkelsen et al. 1987; gerbil: Møller et al. 1985; Shiotani et al. 1986; rabbit, cat, pig: Uddman
et al. 1980; sheep: Cozzi et al. 1989), luteinizing hormone releasing hormone (LHRH;
dog: Matsuura et al. 1983, squirrel monkey: Barry 1979), calcitonin gene-related
peptide (gerbil: Shiotani et al. 1986), vasopressin (VP) and oxytocin (OT; hedgehog:
Nürnberger and Korf 1981; rat: Buijs and Pévet 1979; dog: Matsuura et al. 1983;
macaque: Ronnekleiv 1988). Such peptidergic fibers and terminals can be found in
both perivascular and intraparenchymal locations. Their number and density varies
with the neuropeptide and also the species of mammal.

The origin of peptidergic nerve fibers innervating the pineal organ is not quite clear.
From experiments combining tracing techniques and immunocytochemistry Shiotani
et al. (1986) concluded that in the gerbil VIP immunoreactive nerve fibers originate
from nerve cells located in the pterygopalatine ganglion, and the substance P and
calcitonin gene-related peptide immunoreactive nerve fibers from the trigeminal
ganglion. However, numerous VIP immunoreactive neurons are also present in the
SCG (Sasek and Zigmond 1989) and the hypothalamic paraventricular nucleus (Mikkelsen and Møller 1988), both of which also project to the mammalian pineal organ.
The perikarya of the substance P immunoreactive nerve fibers may be located in the
habenula. According to immunocytochemical investigations, retrograde and anterograde tracing experiments VP and OT immunoreactive nerve fibers appear to originate
from neurons of the paraventricular nucleus of the hypothalamus (Korf and Wagner
1980; Nürnberger and Korf 1981; Larsen et al. 1991). As shown for the hedgehog, the
intensity of the OT and VP immunoreaction exhibited a seasonal variation (Nürnberger and Korf 1981).

Several studies point toward the possibility that also intrapineal cells express neuropeptides and use them for paracrine and/or synaptic signaling. Enkephalin immunoreaction has been shown in neurons of the human pineal gland (Moore and Sibony 1988). In the European hamster selected cells which apparently represent pinealocytes have been shown to be Leu- and Met-enkephalin immunoreactive and to give rise to processes that terminate in club-shaped swellings, occasionally making synapselike contacts with other pinealocytes (Coto-Montes et al. 1994). The idea that cells within the pineal organ express neuropeptides is supported by the demonstration of mRNAs encoding for prepro-enkephalin A (Aloyo 1991) and somatostatin (Mato et al. 1993).

4 Receptor Mechanisms and Second Messenger Systems Involved in the Regulation of the Melatonin Biosynthesis

4.1
General Aspects

According to current concepts, the highly lipophilic melatonin is not stored within the pinealocytes but is released into pineal capillaries immediately after its formation. Thus the secretion of melatonin appears to be controlled solely by its biosynthesis, which is catalyzed by four distinct enzymes: TPH, aromatic *L*-amino acid decarboxylase, NAT, and HIOMT (Fig. 4). In all vertebrate species studied so far, melatonin reaches its maximal concentration during the night and is nearly undetectable during daytime. This rhythm has been shown to depend on the large increase in concentration and activity of the NAT that occurs night by night and increases the concentration of *N*-acetylserotonin which accelerates melatonin synthesis through a mass action effect on HIOMT (Fig. 4). Thus NAT appears as the rate-limiting enzyme of the melatonin biosynthesis and as a molecular interface at which all inputs regulating melatonin production and secretion converge. These inputs are provided by endogenous (circadian) oscillators and photoreceptors.

The photoreceptor signals either influence the endogenous oscillator or have direct effects on the melatonin biosynthesis. In anamniotes the photoreceptor cells controlling melatonin biosynthesis are located in the pineal organ itself. In some species the pineal also contains an endogenous oscillator that controls the melatonin synthesis, and that receives input from the pineal photoreceptors. In mammals the corresponding photoreceptor(s) and endogenous oscillator(s) are located outside the pineal. Light stimuli are perceived by photoreceptors in the retina; the endogenous oscillator is located within the suprachiasmatic nucleus. Thus the rhythm in the melatonin synthesis in the mammalian pineal organ depends on extrapineal signals which are transmitted through neuronal connections (Fig. 3). The major final pathway for the transmission of these signals is the sympathetic innervation with adrenergic receptor and transduction mechanisms. The avian pineal organ represents an intermediate form that contains endogenous oscillators and employs both photoreception/transduction and adrenoreception/transduction as regulatory processes.

In all vertebrate species investigated thus far cAMP and calcium ions (Ca^{2+}) have been identified as the second messengers that stimulate the melatonin biosynthesis. In most cases Ca^{2+} appears to exert its stimulatory effects via an activation of the cAMP pathway. The significance of cGMP for the melatonin biosynthesis has not yet been established, although this second messenger is regulated in a manner very similar to cAMP. It may be that cGMP opens cyclic nucleotide gated cation channels, thereby elevating the concentration of intracellular Ca^{2+} ($[Ca^{2+}]_i$; Schaad et al. 1995).

4.2
Anamniotes

The signal transduction pathways regulating melatonin biosynthesis and release in poikilothermic vertebrates have been identified only partially, but it is clear from all experiments that Ca^{2+} and cAMP play a major regulatory role as second messengers for melatonin biosynthesis. Calcium imaging of isolated and immunocytochemically identified trout pinealocytes reveals that 30% of the pinealocytes exhibit spontaneous $[Ca^{2+}]_i$ oscillations, whose frequency differs from cell to cell (Kroeber et al. 1997; Fig. 21A,B). Neither illumination with bright light nor dark adaptation of the cells has an apparent effect on the oscillations. These observations suggest that the $[Ca^{2+}]_i$ oscillations are independent of the lighting conditions and are not evoked by light-induced activation of phototransduction processes. The trigger for the $[Ca^{2+}]_i$ oscillations remains unknown. They depend on the influx of extracellular calcium through dihydropyridine-sensitive voltage-gated calcium channels that are obviously opened and closed in a rhythmic manner because removal of extracellular Ca^{2+} or application of 10 μM nifedipine causes a reversible breakdown of these oscillations (Fig. 21B). Application of KCl elevates $[Ca^{2+}]_i$ in 90% of the oscillating and 50% of the nonoscillating pinealocytes (Fig. 21C). The effect of KCl is blocked by 50 μM nifedipine. These results suggest that voltage-gated L-type calcium channels play a major role in the regulation of $[Ca^{2+}]_i$ in trout pinealocytes. Notably, comparatively high concentrations of nifedipine are needed to cause complete inhibition of KCl-evoked Ca^{2+} influx. This may indicate that teleosts possess L-type calcium channels somewhat different from those of birds and mammals.

Experiments with thapsigargin reveal the presence of intracellular calcium stores in 80% of the cells, but their physiological role for the regulation of $[Ca^{2+}]_i$ remains elusive. Both calcium stores and L-type calcium channels are also present in avian and mammalian pinealocytes (see below), but their regulation varies from one class of vertebrates to the other.

The majority of S-antigen immunoreactive trout pinealocytes (70%) do not display spontaneous $[Ca^{2+}]_i$ oscillations (Fig. 21C). It is possible that these cells represent a distinct type of pinealocyte, but no correlation can be made between morphological features and the presence of $[Ca^{2+}]_i$ oscillations. Under in vivo conditions $[Ca^{2+}]_i$ oscillations may be transmitted from oscillating to nonoscillating cells via gap junctions.

From a phylogenetic point of view it is interesting to note that NE has no apparent effect on $[Ca^{2+}]_i$ in any trout pinealocyte, regardless of whether the cells display $[Ca^{2+}]_i$ oscillations (Fig. 21A,C) or not. NE also does not influence cAMP levels in trout pinealocytes (Thibault et al. 1993).

Experiments with cultured trout pineal organs show that the melatonin secretion follows a robust, irradiance-dependent secretion profile. Attenuation of light irradiance is associated with a distinct increase in melatonin production whereas enhancement of the illumination is followed by a clear decrease (Fig. 21D; Meissl et al. 1996). In the same type of preparation elevation of the intracellular cAMP concentration have been demonstrated to stimulate NAT activity and melatonin synthesis (Thibault et al. 1993). Increases in intracellular cGMP levels affect the melatonin synthesis only marginally. By contrast, calcium appears as an important mediator of the darkness-induced increase in melatonin production: elevation of $[Ca^{2+}]_i$ enhances melatonin

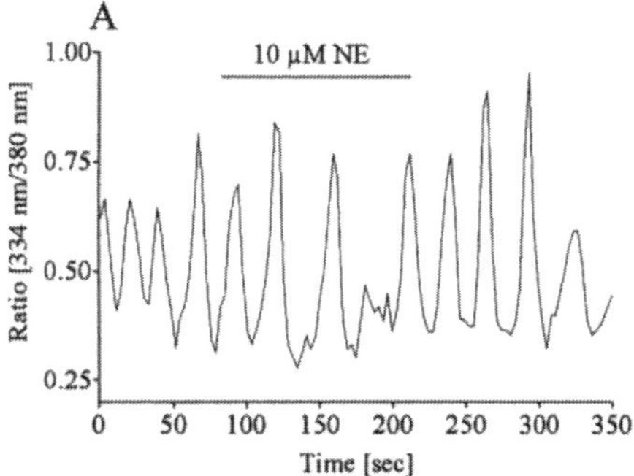
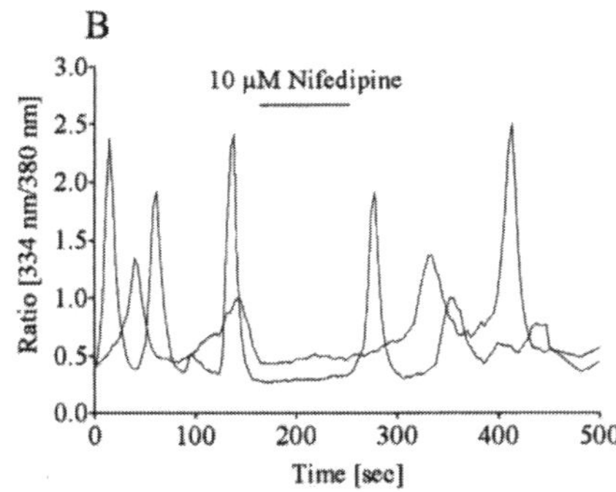
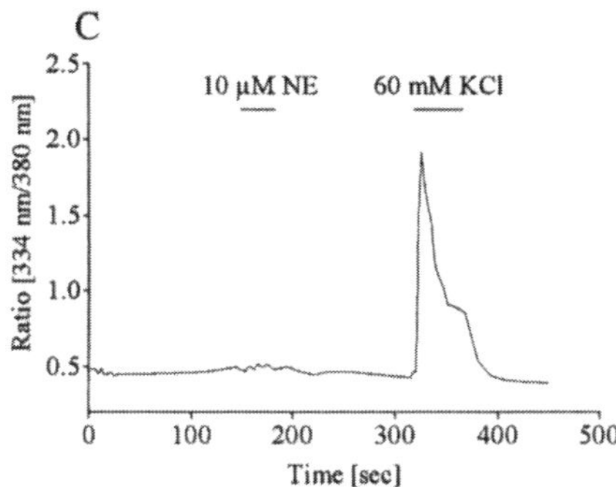
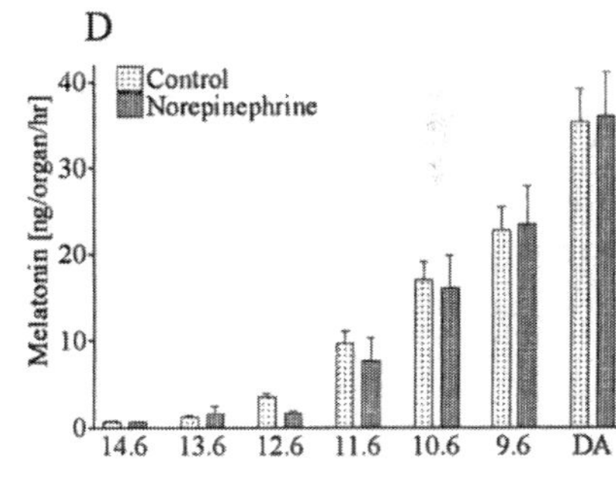

Fig. 21A–D. Regulation of the intracellular calcium ion concentration $[Ca^{2+}]_i$ in S-antigen immunoreactive trout pinealocytes and control of melatonin biosynthesis in the trout pineal organ. **A,B** Several S-antigen immunoreactive trout pinealocytes (up to 30% of the total population) displayed spontaneous $[Ca^{2+}]_i$ oscillations of varying frequencies. **A** These oscillations were not affected by treatment of the cells with 10 μM norepinephrine. **B** Inhibition of voltage-gated calcium channels of the L-type with 10 μM nifedipine blocked the spontaneous $[Ca^{2+}]_i$ oscillations. After removal of the drug the $[Ca^{2+}]_i$ oscillations started again. **C** The majority of the S-antigen immunoreactive trout pinealocytes did not show spontaneous $[Ca^{2+}]_i$ oscillations but had stable basal $[Ca^{2+}]_i$ levels. Also in these cells whose excitability was proven by depolarizing them with 60 mM KCl norepinephrine did not influence the $[Ca^{2+}]_i$ levels. **A–C** adapted from Kroeber et al. 1997; *bars*, duration of the drug treatment. **D** Control of melatonin release from isolated superfused trout pineal organs. The release of melatonin depends on the illumination (light intensity varying from 14.6 to 9.6 log photons $cm^{-2}\,s^{-1}$) and is maximal in dark-adapted (*DA*) organs. Norepinephrine does not influence melatonin secretion. Mean+SEM of n=8 (controls) or n=4 (norepinephrine). (Courtesy of H. Meissl, Bad Nauheim; see Meissl et al. 1996)

production, whereas blocking of calcium channels by nifedipine, verapamil, or cobalt chloride abolishes this increase, probably by the action of these substances on voltage-gated L-type channels (Bégay et al. 1994; Meissl et al. 1996; Kroeber et al. 1997). The elevation of $[Ca^{2+}]_i$ may influence melatonin biosynthesis through its interaction with the cAMP pathway or through other mechanisms which do not involve the cAMP pathway (Bégay et al. 1994).

NE, adrenergic agonists, and dopamine affect neither the maximal release rate of melatonin in darkness nor the melatonin response in the photopic or mesopic range of illumination (Meissl et al. 1996; Fig. 21D), although these catecholamines appear to have a weak modulatory effect on the electrical (neuronal) responses of the trout pineal organ. They enhance the spike rate of pineal second-order neurons but only in relatively high concentrations (Martin and Meissl 1992). The findings from the trout pineal organ suggest that the electrophysiological effects of the catecholamines are functionally separated from the melatonin pathway. On the other hand, catecholamines appear to be involved in the control of melatonin production in the pike pineal organ (Falcón et al. 1991). This indicates possible species differences among teleosts.

Species differences are also evident with regard to the presence of an endogenous oscillator within the pineal organ. The pineal complex of the lamprey, one of the most basic vertebrates, contains an endogenous oscillator (Morita et al. 1992). Some teleost species, for example, the zebra fish and goldfish (Cahill 1996), also show a persisting melatonin rhythm in constant darkness, whereas in salmonids the level of melatonin synthesis is directly related to the level of the ambient illumination (Gern and Greenhouse 1988; Gern et al. 1992; Max and Menaker 1992; Meissl and Yáñez 1996; Zachmann et al. 1992). In general, the melatonin rhythms tend to be damped in the absence of a light-dark cycle. One possible explanation for this phenomenon is that the overt rhythm is the result of an interaction of uncoupled circadian oscillators.

4.3
Birds

As shown by studies with the chicken and the house sparrow, biochemical processes related to the melatonin biosynthesis appear to be regulated primarily by photoreceptor cells and endogenous oscillators that reside in the pineal organ itself. The first experimental evidence for this concept came from investigations by Binkley et al. (1978), Deguchi (1979, 1981), and Takahashi et al. (1980). Subsequently these results have been confirmed and extended by Zatz and coworkers (1988) and Takahashi and coworkers (1989). Chicken pineal organs kept in vitro and deprived of any neuronal connections are able to maintain an endogenous (circadian) rhythm in melatonin production for at least several days (Zatz et al. 1988; Robertson and Takahashi 1988). Moreover, discrete light pulses cause an acute inhibition of the melatonin biosynthesis and entrain the endogenous oscillator in cultured chicken pineal organs. The acute inhibitory effect of melatonin appears to be mediated by cAMP, whereas the entraining effect of light is not (Zatz and Mullen 1988a,b). Cyclic nucleotides (cAMP, cGMP) oscillate in chicken pineal organs kept in vitro with high values during the scotophase and low values during the day. Selective manipulations of cAMP levels without affecting cGMP levels suggest that, as in all other vertebrates investigated to date, cAMP rather than cGMP is the cyclic nucleotide second messenger involved in regulation of the melatonin synthesis.

Calcium also plays an important role for the regulation of melatonin synthesis and secretion in birds. The nocturnal increase in melatonin secretion is suppressed by organic and inorganic Ca^{2+} channel blockers (Harrison and Zatz 1989; Takahashi et al. 1989; Zatz 1992). Voltage-dependent Ca^{2+} channels have been observed in electrophysiological experiments with chicken pineal cells (Harrison and Zatz 1989; Henderson and Dryer 1992). There is evidence that Ca^{2+} acts upon the cAMP pathway (see Takahashi et al. 1989, for review and references). Thus the dose-response curve for 8-Br-cAMP is shifted toward higher concentrations in Ca^{2+}-free external saline, and the nocturnal increase in melatonin production is inhibited by calmodulin antagonists. Influx of Ca^{2+} appears to regulate cAMP levels (Zatz 1992). Together these results suggest that cAMP and Ca^{2+} interact to regulate the activity of NAT and the melatonin secretion. The effects of Ca^{2+} appear to be mediated by calmodulin.

Calcium-imaging in isolated chicken pinealocytes (D'Souza and Dryer 1994) revealed that approximately 10% of isolated chicken pineal cells (which, however, were not identified by immunocytochemical techniques) exhibit spontaneous $[Ca^{2+}]_i$ oscillations. The pools required for the maintenance of these oscillations were not identified in that study, but it may well be that, similar to the results in the trout (Kroeber et al. 1997; see above), voltage-dependent Ca^{2+} channels are involved. Thapsigargin evokes an increase in $[Ca^{2+}]_i$ via mobilization of Ca^{2+} from intracellular stores. Interestingly VIP also induces an increase in $[Ca^{2+}]_i$ in a subpopulation of the cells; in some of these VIP also evokes $[Ca^{2+}]_i$ oscillations. As seen in the trout, NE has no effect on $[Ca^{2+}]_i$ in any quiescent or oscillating cell. Agents that increase intracellular cyclic nucleotides cause mobilization of Ca^{2+}. This may occur through cyclic nucleotide-gated channels that belong either to the rod type (Dryer and Henderson 1991) or to the cone type (Bönigk et al. 1996). Because of its high Ca^{2+} permeability the cone cyclic nucleotide-gated channel may be important for Ca^{2+} entry into the cell.

A novel and fascinating control mechanism for the regulation of $[Ca^{2+}]_i$ in chicken pinealocytes was discovered by D'Souza and Dryer in 1996. Using the patch-clamp technique these authors described a new class of cationic channel that has an unusually long open time and is therefore termed I_{LOT}. This channel is nonselective for cations and does not appear to be gated by voltage or soluble second messengers. The channel is active in chicken pinealocytes at night but not during the day. Interestingly, the rhythm in channel activity persists in pinealocytes kept in constant darkness and is thus under control of the circadian oscillator present in chicken pinealocytes. Blockage of protein synthesis prevents the circadian rhythm in channel activity, indicating that it is regulated through transcriptional and translational control mechanisms rather than through posttranslational modifications. The identification of this channel allows the following hypothesis about regulation of melatonin biosynthesis in chicken pinealocytes. Because the channel is nonselective for cations, pinealocytes are more depolarized at night, and Ca^{2+} can enter the cells through both voltage-sensitive calcium channels and I_{LOT} itself. The elevation of $[Ca^{2+}]_i$ then causes an elevation in intracellular cAMP via activation of the Ca^{2+}/calmodulin-stimulated adenylate cyclase, and the joint increase in Ca^{2+} and cAMP optimally stimulates melatonin synthesis at night (see Takahashi 1996).

The sympathetic innervation and its primary neurotransmitter NE appear to modulate the inherent rhythmicity of the avian pineal organ. In contrast to mammals, NE inhibits the pineal melatonin biosynthesis in birds (see Takahashi et al. 1989, for review and references). The opposite effects of NE on the melatonin biosynthesis in birds and mammals can be readily explained by the fact that the receptors activated by NE differ in the avian and mammalian pineal. Whereas the prevailing adrenoreceptors in the mammalian pineal are α_1- and β-receptors, the chicken pineal organ has α_2-adrenoreceptors (Voisin and Collin 1986). These receptors are coupled to the GTP-binding protein G_i; thus their activation causes an inhibition of the adenylate cyclase activity and a drop in intracellular cAMP (Zatz and Mullen 1988a; Pratt and Takahashi 1987), finally inhibiting the melatonin synthesis.

VIP markedly stimulates melatonin production in the avian pineal organ through elevation of cAMP levels (Takahashi et al. 1989). Obviously this effect is mediated via two VIP binding sites (Meunier et al. 1991). VIP does not seem to interact with NE in the chicken pineal organ because VIP shows no α_1-adrenergic potentiation (Zatz et al. 1990). This is in contrast to the situation in the rat pineal organ.

4.4
Mammals

4.4.1
Adrenergic Receptor Mechanisms

In mammals NE can be considered as the most prominent neurotransmitter conveying signals from the photoreceptors in the retina and the endogenous oscillator in the suprachiasmatic nucleus to the pineal organ and thus appears as the essential regulator of the melatonin biosynthesis. Microdialysis experiments with the rat pineal organ have shown that NE is indeed released in large quantities after the onset of darkness (Drijfhout et al. 1996a). NE acts upon α_1- and β-adrenergic receptors in the pinealocyte membrane (Fig. 22; see Klein 1985, for review and references).

In all mammalian species, including man, activation of β_1-receptors is an absolute requirement for the stimulation of the melatonin biosynthesis during darkness (Axelrod 1974; Klein 1985; Reiter 1991). Activation of β_1-receptors leads to an increase in adenylate cyclase activity, resulting in a tenfold increase in the concentration of intracellular cyclic AMP, which is the essential intracellular second messenger required for the rise in nocturnal melatonin production (Axelrod 1974; Klein et al. 1981). Activation of the β-receptor is mediated via a GTP-binding protein since it can be mimicked by treatment with cholera toxin which acts via ribosylation of G proteins (Ho et al. 1987; Sugden and Klein 1987). Accordingly, the α-subunit of G_s has been found in the pineal organ in relatively high concentrations (Babila et al. 1992). This is consistent with the demonstration of high concentrations of mRNA for $G_s\alpha$ in the pineal organ by means of in situ hybridization (Fig. 23). Neither the protein nor its mRNA displays a diurnal variation.

Stimulation of α_1-adrenergic receptors has no effect on cAMP accumulation on its own, but potentiates the β-adrenergic effect on cAMP accumulation: simultaneous activation of α_1- and β-adrenergic receptors causes a 100-fold increase in cAMP. Since α_1-receptor stimulation also augments cAMP production induced by forskolin (which stimulates the adenylate cyclase directly) or by cholera toxin (which activates G_s), the interactive site for the α_1-and β-adrenergic mechanism appears distal to the β-adrenergic receptor (Chik and Ho 1990). The potentiation is based on a dramatic rise in $[Ca^{2+}]_i$ induced by α_1-adrenergic stimulation (Vanecek et al. 1985). The NE-induced calcium response has been further characterized by the FURA2 technique and image analysis of isolated rat pinealocytes (Schomerus et al. 1995; Korf et al. 1996; Fig. 24). This experimental design shows NE to evoke a dose-dependent, biphasic rise in $[Ca^{2+}]_i$ in more than 95% of the pinealocytes identified by means of S-antigen immunoreac-

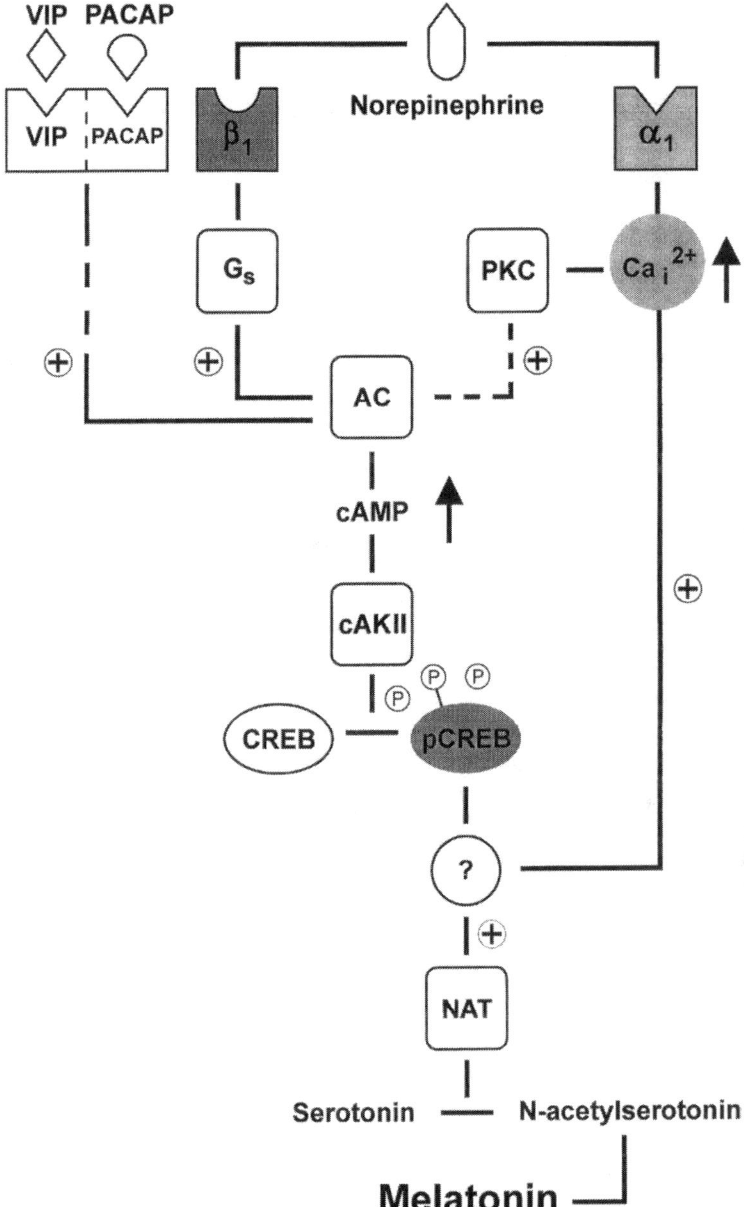

Fig. 22. Schematic drawing of the signal transduction cascade in the mammalian (rat) pineal organ activated upon stimulation of α_1- and β_1-adrenergic receptors or receptors for vasoactive intestinal peptide (*VIP*) or pituitary adenylate cyclase-activating polypeptide (*PACAP*). G_s stimulatory GTP-binding protein; *PKC*, protein kinase C; Ca^{2+}_i, intracellular concentration of free clacium ions; *AC*, adenylate cyclase; *cAMP*, cyclic 3'5'adenosine monophosphate; *cAKII*, protein kinase A type II; *CREB*, cyclic AMP response element binding protein; *pCREB*, phosphorylated CREB; *?*, unknown molecular events; *NAT*, serotonin (arylalkylamine) *N*-acetyltransferase. (Modified and extended after Klein et al. 1992)

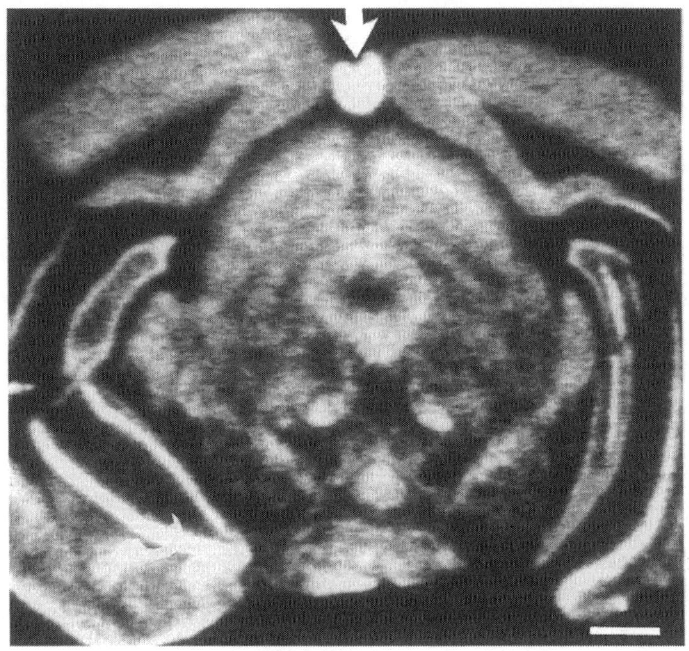

Fig. 23. Autoradiographic demonstration of mRNA encoding for the α-subunit of the stimulatory GTP-binding protein G_s in a frontal frozen section of the rat brain by means of in situ hybridization histochemistry. The pineal (*arrow*) displays the most prominent signal out of all areas included in this frontal section. ×9

tion (Fig. 24A,C). Both the responses and the basal $[Ca^{2+}]_i$ (approximately 50 ± 25 nM) were rather uniform, although the cells showed a conspicuous variation in the intensity of the S-antigen immunoreaction. The responses were confirmed to be mediated by stimulation of α_1-adrenergic receptors because the effects of NE were completely mimicked by phenylephrine, an α_1-adrenergic agonist. The phenylephrine effect was blocked by prazosine, an α_1-adrenergic antagonist, and no changes in $[Ca^{2+}]_i$ were observed after treatment of the cells with the β-adrenergic agonist isoproterenol (Fig. 24A).

The primary phase of the response is characterized by a rapid and transient tenfold increase in $[Ca^{2+}]_i$. This initial peak is followed by a secondary phase in which $[Ca^{2+}]_i$ decreases to a plateau level which is well above the basal $[Ca^{2+}]_i$. The secondary (plateau) phase of the response persists as long as the cells are exposed to NE. Upon removal of the stimulus $[Ca^{2+}]_i$ drops almost immediately to basal levels, and the pinealocytes regain responsiveness to further NE stimulation. The primary phase reflects mobilization of calcium ions from intracellular stores because the initial peak persists when pinealocytes are kept in Ca^{2+}-free medium and then stimulated with NE. In contrast, the secondary (plateau) phase is abolished under these conditions, indicating that it involves an increase in Ca^{2+} influx. Pharmacological manipulations with caffeine and thapsigargin reveal that α_1-adrenergic stimulation activates inositol 1,4,5-trisphosphate-sensitive calcium stores. Depletion of these stores is necessary and also sufficient for the subsequent activation of the Ca^{2+} influx (Schomerus et al. 1995). These data suggest the presence of a store-operated calcium entry that connects phosphoinositide-linked calcium signaling with Ca^{2+} influx. Evidence for a store-operated Ca^{2+} influx has also been obtained in avian pinealocytes (D'Souza and Dryer

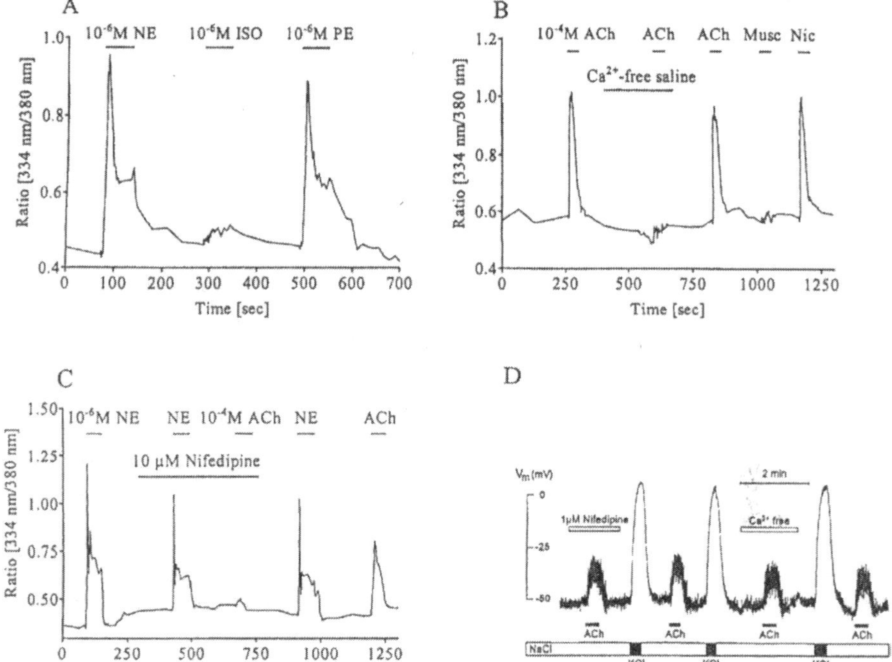

Fig. 24A–D. Regulation of $[Ca^{2+}]_i$ and membrane potential of pinealocytes of adult rats as shown by the FURA-2 method (A–C) or patch-clamp recordings (D). A Norepinephrine (*NE*) elicits a biphasis rise in $[Ca^{2+}]_i$ in virtually all S-antigen immunoreactive rat pinealocytes. The response to the α_1-adrenergic agonist phenylephrine (*PE*) is very similar to the NE-response, whereas the β_1-adrenergic agonist isoproterenol (*ISO*) is without effect. B Acetylcholine (*ACh*) causes a rise in $[Ca^{2+}]_i$ in virtually all S-antigen immunoreactive rat pinealocytes. This response is mimicked by nicotine (*Nic*) and abolished when the cells are kept in calcium-free medium. Muscarine (*Musc*) does not influence $[Ca^{2+}]_i$. C Blocking voltage-gated calcium channels of the *L*-type has very little effect on the calcium response elicited by norepinephrine, but abolishes the ACh-induced response. (For details, see Schomerus et al. 1995) D ACh-induced depolarization of rat pinealocyte which is not significantly reduced in the presence of nifedipine or in the absence of extracellular calcium. A continuous recording of the resting membrane voltage (V_m) is shown during which ACh (100 μM) was applied several times. Bath solution was repeatedly changed from NaCl to KCl solution as indicated below the trace. (For details, see Letz et al. 1997). *Bars*, duration of drug treatments

1994). The channel responsible for this type of influx has not been identified yet. Voltage-gated calcium channels of the *L*-type are present in rat pinealocytes (see below), but they do not appear to be involved in the NE-induced calcium signaling since specific blockers of the *L*-type channel, for example, nifedipine, do not affect the NE-induced signal (Fig. 24C).

The finding that NE causes an elevation of $[Ca^{2+}]_i$ in more than 95% of the immunocytochemically identified pinealocytes via activation of α_1-adrenergic receptors conforms to previous studies showing that calcium plays an important role in mediating the stimulatory effects of NE on the melatonin biosynthesis. Moreover, the data from the single-cell analyses demonstrate that pinealocytes respond to NE stimulation in a rather homogeneous pattern, irrespective of their varying intensities in S-antigen immunoreactivity.

Adrenergic stimulation causes accumulation not only of cAMP but also of cGMP as a consequence of adrenergic stimulation of guanylate cyclase (Klein 1985). This depends on the diffusible messenger nitric oxide (Spessert et al. 1993; Schaad et al. 1994). The precise functional role of cGMP in regulating melatonin biosynthesis remains unknown. It may activate cGMP-gated cation channels, thus leading to an elevation of $[Ca^{2+}]_i$ via an increased Ca^{2+} influx (Schaad et al. 1995).

The density of the β-adrenergic receptors displays a clear 24-h rhythm; its peak occurs in the late light phase (Romero et al. 1975; for species other than the rat, see Reiter 1991). After superior cervical ganglionectomy an increased binding of β-agonists to the pineal is observed (Cantor et al. 1981). Furthermore, ganglionectomy destroys the daily rhythm in pineal β-receptor density. Thus NE is obviously involved in regulating the diurnal rhythm in the number of β-receptors and also in that of α_1-receptors (see Reiter 1991, for review and references). In situ hybridization and northern blotting have shown diurnal variation in the mRNA encoding for the β-receptor (Carter 1993; Møller et al. 1997; Stehle et al. 1997), but the peak in mRNA levels occurs in mid-dark (at 24:00 hours) whereas lowest level is seen at midlight (12:00 hours). Bilateral removal of the SCG decreases the amount of mRNA in the pineal. The considerable time lag between the maximum level of mRNA and the maximum ligand binding indicates that de novo synthesis of the receptor protein is not involved in the acute up- and downregulation of the receptor (see also p. 64).

4.4.2
Receptors and Binding Sites for Neuropeptides, Serotonin, Acetylcholine, and Glutamate

In addition to adrenoreceptors, the mammalian pineal organ is endowed with highly specific binding sites for several neuropeptides, neurotransmitters, and drugs (see Ebadi and Govitrapong 1986; Ebadi et al. 1989; Cardinali et al. 1987). These include bindings sites for VIP, peptide N-terminal histidine and C-terminal isoleucine (PHI), pituitary adenylate cyclase-activating peptide (PACAP), NPY, glutamate, GABA, benzodiazepine, dopamine, ACh, serotonin, delta-sleep inducing peptide, substance P, di-ortho-tolylguanidine (so called sigma-receptors, Abreu and Sugden 1990; Jansen et al. 1990), and adenosine (Sarda et al. 1989; Stehle 1995). The role of these substances for the regulation of pineal functions has been elucidated only to a limited extent. Nearly all studies related to this topic have addressed the questions of whether and how these neurotransmitters, neuropeptides, and drugs influence the melatonin biosynthesis.

4.4.2.1
VIP, PHI, PACAP

The most convincing results to date have been obtained with VIP. The presence of receptors for VIP has been shown in the rat and gerbil pineal gland by means of biochemical and autoradiographical methods (Kaku et al. 1985; Møller et al. 1985). Scatchard analyses indicate the presence of two classes of binding sites: high-affinity, low-capacity sites and low-affinity, high-capacity sites (Kaku et al. 1985). These are independent of the β-adrenoreceptors because preincubation with NE does not influence the VIP binding on the pinealocytes (Kaku et al. 1985). Treatment of the animals with constant light increases the number of both VIP receptor types (Kaku et al. 1985). This may be attributed to a supersensitivity of the receptors which results from an inhibition of VIP release from nerve terminals during the photophase (Kaku et al. 1986). The receptors thus seem to be subject to up- and downregulatory mechanisms. In the rat pineal VIP elevates both basal and NE-stimulated NAT activity (Yuwiler 1983) and causes accumulation of both cAMP and cGMP (Kaneko et al. 1980; Chik et al. 1988). The finding that VIP-induced increase in NAT is blocked neither by superior cervical ganglionectomy nor by pretreatment with reserpine supports the notion that VIP exerts its effect via "postsynaptic" receptors on the pinealocyte membrane (Yuwiler 1983). α-Adrenergic agonists potentiate the effects of VIP with regard to both parameters: activation of NAT (Yuwiler 1987) and accumulation of cGMP and cAMP (Chik et al. 1988).

The issue of whether VIP also influences $[Ca^{2+}]_i$ has not yet been clarified. In our own laboratory we have not been able to record a calcium response in rat pinealocytes upon stimulation with VIP. On the other hand, Schaad et al. (1995) have shown that VIP induces a moderate increase in $[Ca^{2+}]_i$ in approximately 65% of rat pinealocytes kept in culture. This rise appeared to be mediated via an increased influx into the pinealocytes through cGMP-gated cation channels, because (a) some of the VIP effects were mimicked by treating the cells with membrane-permeable cGMP analogs, and (b) the VIP-induced rise in $[Ca^{2+}]_i$ was blocked by *L*-cis diltiazem that specifically inhibits this type of channel (Schaad et al. 1995). As in rat pinealocytes, VIP has been found to stimulate cAMP accumulation in ovine pinealocytes (Morgan et al. 1988, 1989).

These findings suggest that VIP substitutes β-adrenergic agonists and interacts with NE by means of independent receptors, and they raise the interesting possibility that the regulation of pineal functions in mammals depends not only on catecholamines but also on neuropeptides. The two neurotransmitter types may interact with each other. Moreover, the results suggest that sympathetic nerve fibers originating from the SCG interact with nerve fibers originating from other sources (i.e., parasympathetic ganglia, nuclei of the di- or mesencephalon).

Another peptide derived from the VIP-precursor molecule, PHI, also stimulates melatonin biosynthesis (Moujir et al. 1992). PHI has been shown to promote cAMP synthesis (Kaneko et al. 1980) via specific receptors coupled to adenylate cyclase (Tsuchiya et al. 1987). Similar results have been obtained with helodermin, which was originally isolated from the venom of the Gila monster lizard, belongs to the secretin family, and is structurally similar to VIP and PHI (Kaku et al. 1992).

Moreover, the pineal gland contains high concentrations of receptors for PACAP (Masuo et al. 1992) which also belongs to the VIP/secretin/glucagon family and exihibits significant homology with VIP (Miyata et al. 1989). Two types of receptors have been identified. Type I receptors specifically bind PACAP38 and PACAP 27 and have low affinity for VIP. Type II receptors bind both PACAP and VIP with comparable affinities. In the rat pineal type I receptors appear to be 12-fold more abundant than type II receptors (Masuo et al. 1992). PACAP has been shown to increase cAMP accumulation 4- to 5-fold, and this effect was potentiated 2- to 3-fold by α_1-adrenergic activation (Chik and Ho 1995). PACAP alone stimulated melatonin production 2.5-fold (Simonneaux et al. 1993; Chik and Ho 1995), and phenylephrine potentiated the

PACAP-induced melatonin production. Interestingly, PACAP applied with or without phenylephrine did not affect cyclic GMP levels. Thus PACAP is the first substance identified that increases cAMP accumulation without increasing cGMP levels. These findings suggest that the VIP receptors present in the pineal gland are distinct from PACAP type II receptors (Chik and Ho 1995).

4.4.2.2
NPY

Relatively few studies have dealt with binding sites for NPY and its influence on pineal functions, despite the fact that NPY immunoreactive nerve fibers are quite abundant in the mammalian pineal organ. The results of these studies remain somewhat equivocal. Injection of NPY into the right carotid artery of intact rats led to an increase in NAT activity. However, when NPY was administered to blinded animals, the neuropeptide provoked a drastic inhibition of the enzyme activity (Reuss and Schröder 1987). Using rat pineal explants, Vacas et al. (1987) showed that NPY augments both the basal and the NE-induced production and release of melatonin. On the other hand, Olcese (1991) investigated suspended rat pinealocytes (devoid of any nerve fibers as presynaptic elements) and found that NPY is a potent inhibitor of the NE-stimulated melatonin secretion. Surprisingly, the inhibitory effect was accompanied by only a modest reduction in the intracellular cAMP level. These divergent results suggest that NPY acts upon both pre- and postsynaptic receptors in a different manner, and that in the presence of presynaptic elements the inhibitory effect of NPY on the pinealocyte membrane itself is masked by the excitatory response mediated via an activation of presynaptic receptors. Investigation with isolated rat pinealocytes kept in culture have shown that pretreatment of the cells with NPY appears to reduce the NE-induced calcium signal in approximately 40% of the pinealocytes (Korf et al. 1996). This effect may be one event responsible for the inhibitory action of NPY on melatonin secretion.

4.4.2.3
Vasopressin/Oxytocin

Divergent results have also been obtained in regard to the effects of VP and OT on the mammalian pineal. Schröder et al. (1988) noted that these neuropeptides inhibit the nocturnal rise in melatonin when injected into rats at the early dark phase, i.e., at a time, when the adrenergic receptor system is activated. Investigating rat pineal glands in vitro, Simonneaux et al. (1990) found that VP and OT decrease the basal melatonin production at rather low concentrations (10^{-6} M or 10^{-7} M) but potentiate the isoproterenol-induced increase in melatonin secretion. In vivo and in vitro investigations with Sprague-Dawley and Brattleboro rats led Stehle et al. (1991) to suggest two mechanisms which mediate the effect of VP on pineal melatonin synthesis. The inhibitory effect observed in vivo may be attributed to an inhibitory action of VP on elements of the pinealopetal innervation, which may mask the potentiating effect of VP on the pinealocytes themselves.

A significant expression of mRNA encoding for the V_{1a} subtype of the VP receptor has been demonstrated in the rat pineal organ (Ostrowski et al. 1994). This receptor type is known to stimulate phosphatidyl inositol turnover and to mobilize Ca^{2+} from intracellular stores. However, calcium imaging reveals that VP causes a rise in $[Ca^{2+}]_i$ only in a minority of rat pineal cells which do not display S-antigen immunoreaction (Schomerus et al. 1995).

4.4.2.4
Serotonin

Serotonin has been shown to amplify the β-adrenergic induction of NAT in cultured
rat pinealocytes but has no effect on enzyme activity when added alone (Sugden 1990).
The receptor type involved in this response may be similar to the 5-HT$_{1P}$ receptor
described in the enteric nervous system. An interesting question is the source of
serotonin. Serotonin is taken up by noradrenergic (sympathetic) nerve endings in the
pineal and may be released upon their stimulation. However, pinealocytes themselves
are more likely to be the source for serotonin. There is some evidence that the loss of
serotonin occurring in these cells soon after darkness (Quay 1974) is due not only to
the induction of the melatonin biosynthesis but also to the release of serotonin into
the surrounding extracellular fluid. Interestingly, this release seems to be under
control of adrenergic receptors (Aloyo and Walker 1987, 1988).

4.4.2.5
Acetylcholine

Possible effects of cholinergic agonists on melatonin production and release are of
peculiar interest because ACh is the primary neurotransmitter candidate of a presumed parasympathetic innervation of the mammalian pineal organ, and an antagonistic interaction between the sympathetic and the parasympathetic system may exist
in the mammalian pineal organ, as is the case with other glands. Cholinergic effects on
pineal function have been repeatedly investigated (see Laitinen et al. 1995), but the
data are equivocal with regard to the receptor types involved, their location, and the
functional consequences of their activation. Transpineal in vivo microdialysis has
shown that infusion of the cholinergic agonists carbachol or oxotremorine into the rat
pineal organ results in a marked decrease in melatonin release during the dark period
by inhibiting the release of NE from the sympathetic nerve fibers in the pineal
(Drijfhout et al. 1996b). Such data suggest the presence of muscarinic receptors in a
presynaptic location, i.e., on the sympathetic nerve fibers in the pineal organ.

The existence of nicotinic ACh receptors has been inferred from immunocytochemical investigations (Reuss et al. 1992) and binding studies with radiolabeled
ligands (Stankov et al. 1993), and it has been suggested that nicotine has an inhibitory
effect on pineal melatonin synthesis. The existence of a nicotinic type of ACh receptor
on rat pinealocytes has also been shown by calcium imaging and patch-clamp investigations (Fig. 24B,D; Schomerus et al. 1995; Letz et al. 1997). Calcium imaging reveals
that more than 90% of isolated, immunocytochemically identified pinealocytes respond to ACh or nicotine in a dose-dependent manner (Fig. 24). All ACh-sensitive cells
also respond to NE, but the two types of responses are quite different. The ACh-induced
elevation of $[Ca^{2+}]_i$ is followed by a rapid decrease to basal levels within a few minutes
after the onset of the stimulus. This decrease is also seen under constant exposure to
ACh. The response is mediated by a nicotinic receptor subtype because the responses
to ACh and nicotine are virtually identical, and both the ACh- and the nicotine-induced
responses are blocked by the specific nicotinic antagonist, d-tubocurarine. Muscarinic
receptor types do not play a role for this calcium response (at least in pinealocytes
obtained from adult rats) because pilocarpine acting upon all presently known muscarinic receptor subtypes does not evoke a calcium response, and the muscarinic
antagonist atropine does not block the ACh-induced rise in $[Ca^{2+}]_i$. The response to

ACh is totally prevented when pinealocytes are kept in Ca^{2+}-free saline, indicating that the cholinergic effects are based upon Ca^{2+} influx.

The components involved in this response have been identified using a combination of patch-clamp recordings and calcium imaging (Letz et al. 1997). These investigations have shown that the resting membrane voltage V_m of isolated rat pinealocytes averages –43 mV, and that replacement of extracellular NaCl by KCl completely depolarizes the cells. This indicates that the resting V_m is dominated by a K^+ conductance. Single-channel recordings reveal the presence of a large conductance Ca^{2+}-activated charybdotoxin-sensitive K^+ channel (Letz et al. 1997; see also Ceña et al. 1991). Application of ACh depolarizes the pinealocytes on an average by 16 mV. The depolarizing effect of ACh is mimicked by nicotine and prevented by tubocurarine. This depolarization is largely abolished in the absence of extracellular Na^+ but is not significantly affected by extracellular Ca^{2+} removal. The ACh-induced rise in $[Ca^{2+}]_i$ is largely reduced after extracellular Na^+ removal. Nifedipine reduces the ACh-induced increase in $[Ca^{2+}]_i$ by approximately 50%. The findings indicate that in rat pinealocytes stimulation of a nicotinic ACh receptor induces depolarization mainly by Na^+ influx through the receptor. The depolarization then activates L-type Ca^{2+} channels which are responsible for the nifedipine-sensitive portion of the intracellular Ca^{2+} increase.

The fact that nicotinic receptors are present in the vast majority of rat pinealocytes suggests its important role in regulation of pineal metabolism. One possible mode of action is that the stimulation of nAChRs and subsequent activation of L-type calcium channels stimulate the release of glutamate via exocytosis of microvesicles in the pinealocytes (see below).

4.4.2.6
Glutamate

The role of glutamate for pineal signaling has received increased interest in recent years. As described above, several immunocytochemical studies (McNulty et al. 1992; Redecker and Veh 1994) have suggested that mammalian pinealocytes employ glutamate as neurotransmitter. Quisqualate-sensitive glutamate binding activity and expression of kainate KA-2 receptor mRNA have been demonstrated in the mammalian pineal (Govitrapong et al. 1986; Kus et al. 1993; Wisden and Seeburg 1993). Moreover, L-glutamate has been shown to inhibit NE-stimulated melatonin biosynthesis (Govitrapong and Ebadi 1988; Van Wyk and Daya 1994; Kus et al. 1994). Recent studies by Yamada et al. (1996a,b) reveal that glutamate is secreted from pinealocytes through microvesicle-mediated exocytosis. This process depends on an increase in Ca^{2+} influx elicited by depolarization of the cells with at least 25 mM KCl and partly mediated by L-type calcium channels. The results of Yamada et al. (1996a,b) on glutamate increases $[Ca^{2+}]_i$ in isolated rat pinealocytes also confirm that glutamate inhibits the NE-induced melatonin secretion.

The data on glutamate and ACh leads to a hypothesis that could explain the inhibitory effect of ACh on the melatonin biosynthesis. Via the nicotinic receptors on the pinealocyte membrane ACh depolarizes the cells, activates L-type calcium channels, and causes an increase in Ca^{2+} influx. This stimulates the release of glutamate from pinealocytes which then inhibits melatonin synthesis and secretion. Further experimental elucidation of this model will help to understand putative antagonistic effects between ACh and NE and to clarify the role of paracrine signaling in the pineal organ which involves glutamate as active substance.

5 Pineal Transcription Factors and Their Possible Roles for the Regulation of the Melatonin Biosynthesis

In recent years considerable progress has been made toward identifying pineal transcription factors and their possible role for the control of melatonin biosynthesis. These studies (nearly all of them performed with mammals) have highlighted the importance of transcription factors that are controlled by the NE/cAMP pathway and impinge on genes bearing a promoter element sensitive to elevation of intracellular cAMP levels (CRE, see below). An essential component of this transmembrane signaling pathway is the cAMP-inducible protein kinase A (PKA). This enzyme consists of catalytic and regulatory subunits (see Maronde et al. 1997a) which are aggregated under resting conditions but dissociate upon cAMP elevation. The catalytic subunits of the PKA phosphorylate constitutively expressed proteins in the cytoplasm and/or – upon translocation into the nucleus – nuclear proteins, as transcription factors. Transcription factors can bind to promotor and enhancer sequences located on genes upstream of the start site of DNA transcription to modulate the activity of DNA polymerases. This leads to an activation or inhibition of DNA transcription and constitutes a fundamental link between an external cue and an adaptive cellular response to appropriately react in time and space.

An important common step in gene targeting is the induction of proto-oncogenes, or so called immediate early genes (IEGs). IEGs are rapidly induced, and their translational products function as transcriptional regulators (Curran et al. 1990). There is increasing evidence for cAMP-dependent signaling events to be mediated via the induction of IEGs through a variety of stimuli (kinases, growth factors, transforming oncogenes, transmitters). Induction of the prototypic and most intensely investigated IEG c-*fos* has been shown to be rapid but transient in neurons in vivo (Sheng and Greenberg 1990; Morgan and Curran 1991). Products of the IEGs c-*fos*, c-*jun*, and *fra-2* form homo- and heterodimeric DNA binding proteins of the activator protein-1 (AP-1) transcriptional regulatory complex that interacts with a common DNA binding site on late response genes, the serum response element (TRE; palindromic consensus sequence: TGACTCA) and accounts for the versatility in the transduction of synaptic signals. AP-1 exerts a substantial transcriptional role in normal cell growth, promotion of the transforming phenotype, and differentiation.

In rat pineal gland a rhythmic transcription has been demonstrated for the IEGs c-*fos*, *jun-B* (Carter 1992), and *fra-2* (fos-related antigen; Baler and Klein 1995), forming variable complexes of homo- and heterodimeric AP-1 protein transcription factors (Carter 1994; Baler and Klein 1995). Bandshift assays of rat pineal extracts have demonstrated a diurnal rhythm in the activity of AP-1, in parallel to melatonin synthesis. This rhythm is mediated via β-adrenergic transmembrane signaling. Variation in the amount of the AP-1 complex seems to be unrelated to the FOS protein but is rather shaped by dimerization of the β-adrenergically regulated IEG JunB with a constitutively expressed component, the IEG JunD. Interestingly, the homodimeric IEG complex FRA-2 shows a similar pattern of expression as AP-1, with levels rising before or in parallel to NAT mRNA increase (see below; Borjigin et al. 1995; Roseboom et al. 1996). In rat and hamster pineal gland a sharp peak in FOS expression has been demonstrated in the middle of the dark period (i.e., 7 h and 5 h after lights off; Koistinaho and Yang 1990). Noteworthy is that the translational product FOS appears in rat pineal gland rather late, i.e., only 5 h after the peak in c-*fos* transcription. This is in contrast to the rapid sequence of these events observed in other systems (Curran et al. 1990). It may be that the FOS-like immunoreactivity observed in the rat pineal gland by Koistinaho and Yang (1990) is elicited by the product of the *fra-2* gene since FRA is also induced in a somehow delayed manner, with peak values not appearing before 5 h after the start of the NE stimulation. Immunoblotting with an antibody that detects both the FOS and the FRA-2 protein has revealed that a 10- to 100-fold lower concentration of NE is sufficient to induce FRA-2 than FOS. β-Adrenergic receptor activation drives the circadian expression of FRA-2 in rat pineal gland, thus resembling regulatory mechanisms characterized for AP-1 (Baler and Klein 1995).

The expression of the two AP-1-forming proteins, FOS and JUN, has been shown to be differentially regulated (Carter 1992, 1993). Induction of the c-*fos* gene is primarily mediated through activation of α_1-adrenergic receptors and the PKC pathway, while c-*jun* is regulated via a dual adrenergic mechanism with opposite effects: the α_1-adrenoceptor/PKC pathway stimulates, while the β_1-adrenoceptor/PKA pathway inhibits, the expression of this IEG. It should, however, be kept in mind that the in vitro data on induction and regulation of pineal IEGs are often not consistent with results obtained in vivo. Moreover, it must be pointed out that there still exists a major gap in information about the functional role of the IEGs characterized in the rat pineal gland so far; it is not yet known whether they function as activator or inhibitor of cAMP-linked gene expression, and their target genes also remain enigmatic.

Still, IEGs that show day/night (diurnal) or oscillator-regulated (circadian) differences in their transcription pattern, expression, structure, and activity properties may in one way or another modulate rhythmic melatonin biosynthesis (Stehle 1995). For example, it may be hypothesized that IEGs are induced upon sudden perturbations of the oscillator (SCN)-regulated melatonin synthesis, thus functioning in acute "emergency" situations. This view is backed by the rapid IEG induction in the SCN which occurs when animals are exposed to light during nighttime (Kornhauser et al. 1990).

The transcriptional control executed by IEGs through AP-1-binding to TRE promoter elements is, however, only one-half of the orchestra which plays on pineal genes. This half must be considered in concert with modulation of gene expression via transcription factors binding to CRE (palindromic consensus sequence: TGACGTCA; Masquilier and Sassone-Corsi 1992). This sequence is detected on an increasing number of genes and characterized as a functional promoter element. The wide distribution and the conserved structure of the CRE suggests that cAMP-mediated gene expression is regulated through a common and possibly widely distributed DNA binding protein, whose activity is modulated by PKA. Indeed, the cloning of CREB from PC12 cells confirms this suggestion to be correct (Montminy and Bilezikjian 1987). CREB is a 43-kDa protein that is ubiquitously expressed (a typical feature of housekeeping genes), functions upon activation as a mediator of a variety of physiological events, and has been shown to be a prerequisite for initiation of immediate/late response gene induction in many systems (Sheng and Greenberg 1990). It is therefore not surprising that CREB mRNA is ubiquitously expressed in the central nervous system including the pineal gland (Stehle et al. 1993). CREB activation is accomplished by phosphorylation of the amino acid serine at position 133 (Gonzales and Montminy 1989). Since phosphorylation induces a conformational change of the CREB protein, a specific antibody for phosphorylated CREB (pCREB) could be generated which selectively detects the activated form of this transcription factor (Ginty et al. 1993).

Recent studies using gel mobility shift assays (Roseboom and Klein 1995) and immunocytochemical and immunochemical detection of pCREB and CREB (Tamotsu et al. 1995; Schomerus et al. 1996; Maronde et al. 1997b; Fig. 25) have substantiated the important role of pCREB as a pineal transcription factor. The stimulation of isolated rat pinealocytes with NE or the β-adrenergic agonist isoproterenol results in a similar, time-dependent induction of pCREB immunoreactivity exclusively located in cell nuclei of almost every pinealocyte (Tamotsu et al. 1995). Nuclear staining shows a maximum after 30–60 min of stimulation (Fig. 25a) and persists for up to 5 h, provided the cells remain exposed to the adrenergic stimulus. Upon removal of NE the pCREB immunoreaction declines to background levels within 60 min (see Korf et al. 1996). α_1-Adrenergic agonists do not induce pCREB in pinealocytes (Fig. 25b) and do not seem to potentiate the isoproterenol-induced phosphorylation of transcription factor CREB. Thus NE-induced phosphorylation of CREB in the pineal organ seems to depend on an activation of the cAMP-PKA pathway. Experiments with agonists and antagonists that distinguish between type I and type II of PKA show that the phospho-rylation of CREB is mediated by PKA type II (Maronde et al. 1997b). Interestingly, these studies also reveal a tight correlation between the activation of PKA type II, the extent of CREB phosphorylation, and the melatonin concentration in the medium of the cells. From the molecular viewpoint these data confirm the hypothesis that, at least in the rat, pineal gene expression is controlled pivotally by the NE/cAMP/PKA pathway centered around the phosphorylation of CREB.

VIP and PACAP, which stimulate the adenylate cyclase and melatonin synthesis via membrane-bound receptors (see p. 54ff), have also been shown to elicit phosphoryla-tion of CREB in pinealocytes (Schomerus et al. 1996), although to a lesser extent and in only 50%–65% of pinealocytes (Fig. 25c), indicating that receptors for VIP and PACAP are expressed only in a subpopulation of pinealocytes. In contrast to repetitive noradrenergic stimulation, a second peptidergic challenge of pinealocytes does not reinduce phosphorylation of CREB. This may indicate a more rapid and effective desensitization of signal transduction mechanisms for VIP and PACAP than for the adrenergic pathway. Notably, VIP and PACAP are also less effective than NE in inducing melatonin biosynthesis and release. The amount of total (phosphorylated and unphosphorylated) CREB is not changed upon stimulation of the cells with NE, VIP, or PACAP (Schomerus et al. 1996; Fig. 26).

We have recently demonstrated a circadian rhythm in CREB phosphorylation in the rat pineal gland in vivo. This rhythm is driven by signals from the endogenous oscillator in the SCN which activate β-adrenergic receptors on pinealocytes (J.H. Stehle, E. Maronde, and H.-W. Korf, unpublished observations). The rhythmic vari-ation in the amount of pCREB in rat pineal gland is tightly correlated to the cAMP-de-pendent formation of melatonin. Quantitative analyses of the pCREB induction in pinealocytes indicate that only a minor proportion of CREB needs to be phosphory-lated to mediate a full stimulation of melatonin synthesis. This fact may explain the

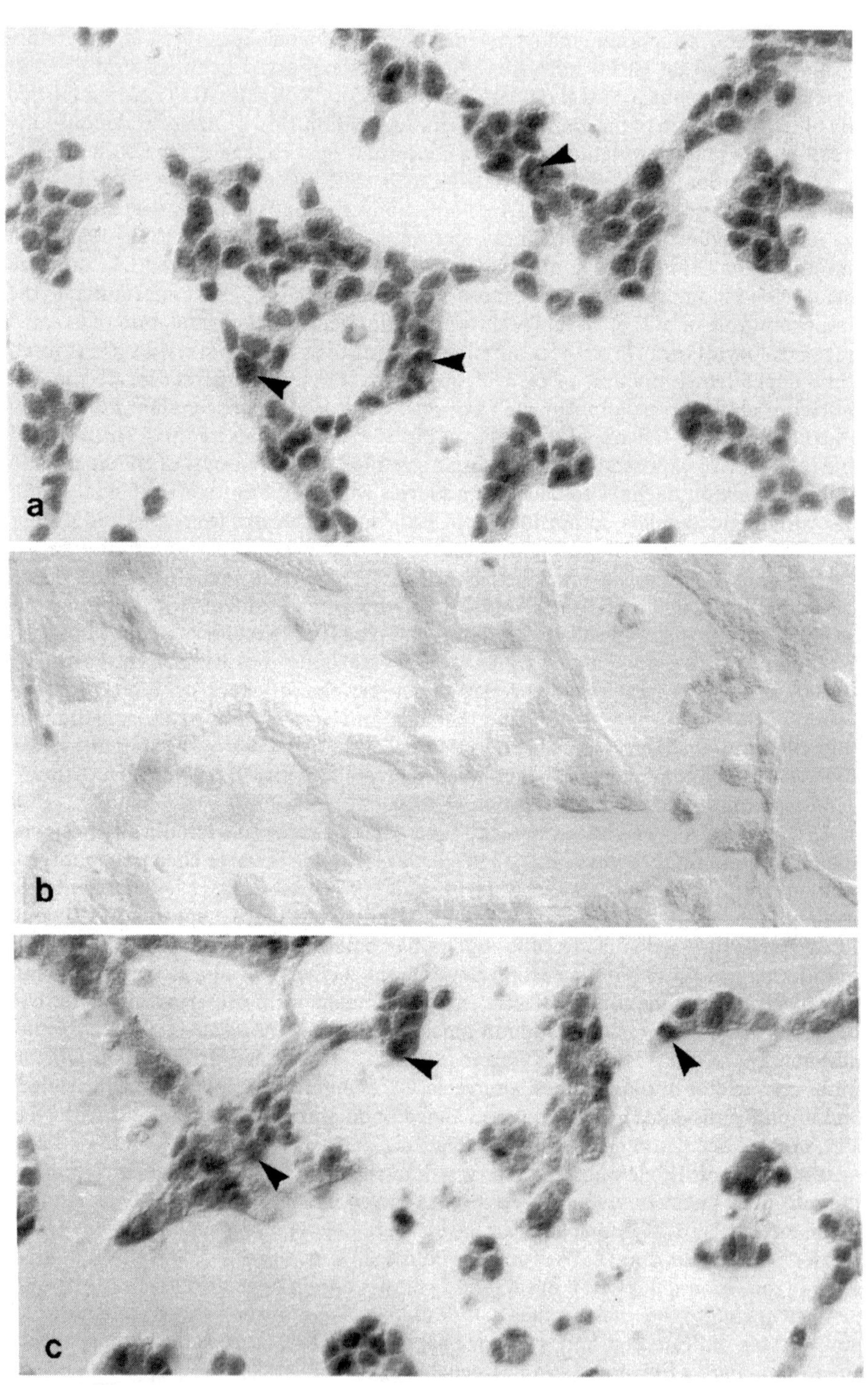

a
b
c

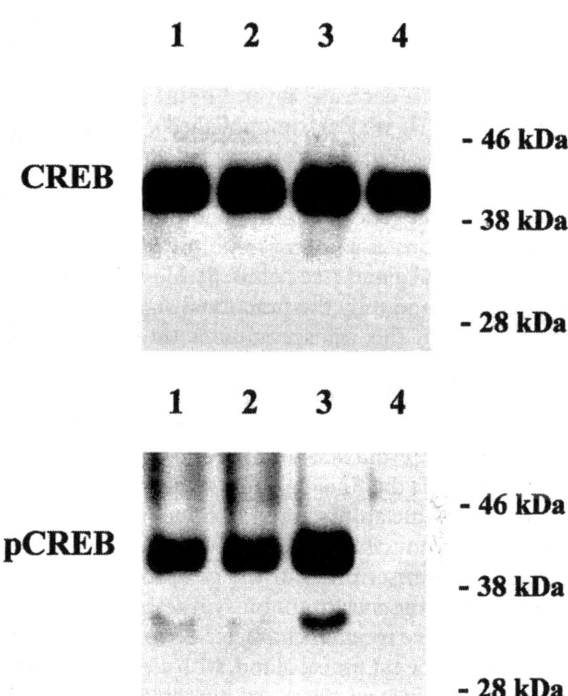

Fig. 26. Immunoblots with antibodies recognizing either total (phosphorylated and unphosphorylated) CREB or its phosphorylated form only (pCREB). On each lane proteins from 35,000 rat pinealocytes were loaded. Cells were treated with 10^{-7} M PACAP 27 (*lane 1*), 10^{-7} M VIP (*lane 2*), or 10^{-6} M norepinephrine (*lane 3*). *Lane 4*, unstimulated controls. The total CREB antibody detects a protein of 42 kDa and two bands at 34 and 30 kDa, the 42-kDa band representing more than 80% of the total signal detected by the pCREB antibody. Calculation of the relative optical density showed that the pCREB signal elicited by norepinephrine is approximately three times stronger than those elicited by PACAP or VIP. (From Schomerus et al. 1996)

notorious difficulties that we and others have encountered in detecting the circadian pCREB rhythm in rat pineal gland. The high amount of "back-up" CREB protein in rat pineal gland may serve to compensate for the low efficiency of Ser-133 phosphorylation of CREB by the catalytic subunit of the PKA (Hagiwara et al. 1993). Additionally, the large pool of total CREB might serve to ensure a broad dynamic range in the phosphorylation of CREB, required by the environmental and physiological situations of the animals. However, a nuclear pCREB signal can always be detected at least 1 h prior to the rise in pineal melatonin synthesis and is a reliable marker for the rapid (within 30 min) stimulus-induced activation of the cAMP-signaling pathway under both in vivo and in vitro conditions (Roseboom and Klein 1995; Tamotsu et al. 1995; Schomerus et al. 1996).

In vivo, pCREB levels stay elevated throughout more than half of the night, indicating the need of a maintained high pool of the activated transcription factor for nocturnally elevated pineal metabolism. In the rat NAT transcription also stays elevated throughout the night (Borjigin et al. 1995; Roseboom et al. 1996; see below), and a causal relationship between CRE-mediated transcriptional activation and the amount of pCREB – already demonstrated in PC12 cells (Hagiwara et al. 1992, 1993) –

Fig. 25a–c. Immunocytochemical demonstration of phosphorylated CREB in nuclei (*arrowheads*) of isolated rat pinealocytes. Response to treatment with norepinephrine (a), VIP (b), and PACAP (c) for 30 min. a–c ×400

appears likely. The mechanisms involved in downregulation of NAT transcription occurring at the end of the night are the focus of intense current research. The level of pCREB starts to decrease several hours before the end of the night (J.H. Stehle, E. Maronde, and H.-W. Korf, unpublished). Thus the potency of this transcription factor to activate gene transcription may fade because of dephosphorylation of CREB. This event is flanked by another transcriptional mechanism, the active downregulation of cAMP-inducible genes by the inducible inhibitory transcription factor ICER. The ICER protein functions as a potent inhibitor of the transcription of cAMP-inducible genes in the rat pineal gland (see below; Stehle et al. 1993, 1995).

To understand fully the functional significance of the CREB phosphorylation, the target genes for this transcription factor must be identified. Several genes encoding molecules of the signal transduction cascade that drives pineal melatonin biosynthesis bear a CRE as a promotor element: (1) the gene for the β_1-adrenergic receptor, (2) the gene for the TPH, the enzyme that converts tryptophan into 5-hydroxytryptophan, (3) the gene for NAT, the rate-limiting enzyme in melatonin biosynthesis, the (4) the gene for the HIOMT, the final enzyme of the melatonin biosynthesis converting N-acetyl-serotonin into melatonin, and (5) the gene for ICER, the potent transcriptional inhibitor of cAMP-inducible genes.

1. Even the transcription of the gene for the β_1-adrenergic receptor, the initiating element of the melatonin biosynthesis, is inducible via the cAMP-signaling pathway. We have recently characterized a circadian rhythm in β_1-adrenergic receptor mRNA in the rat pineal gland, with elevated levels linked to the intrapineal release of NE (Stehle et al. 1997). An upregulation in receptor mRNA occurs shortly after the beginning of the dark period, and an endogenous downregulation can be observed before the end of the night. β_1-Adrenergic receptor transcription is not directly correlated with receptor protein levels since the number of ligand binding sites (see Pangerl et al. 1990) shows elevated levels at the earliest 7 h after the nocturnal peak in β_1-adrenergic receptor mRNA. This suggests that the receptor protein rests in a sequestered state, and that its incorporation into the pinealocyte membrane is suspended until the lack of ligand during daytime triggers this event. The pineal β_1-adrenergic receptor would thus obey classical ligand-dependent regulations common for many other G protein coupled receptors (Hausdorff et al. 1990); rhythms in β_1-adrenergic receptor mRNA and protein are out of frame and driven either by the presence (mRNA) or the absence (protein) of ligand. Thus the biochemical activities in the rat pineal gland, induced by the nocturnally elevated release of NE, seem to be stabilized in two ways even at the β_1-adrenergic receptor level: NE induces transcription of the β_1-adrenergic receptor, thus providing a basis to react to daytime ligand deprivation with an upregulation in receptor abundancy. At the same time NE rapidly elicits sequestration of ligand binding sites as a protective mechanism to diminish physiological responses in the presence of a constant NE stimulus. As the night progresses, activation of the pineal cAMP-signaling pathway is maintained to promote CREB phosphorylation and transcription of cAMP-inducible pineal genes such as NAT (see below; Borjigin et al. 1995; Roseboom et al. 1996), ICER (see below; Stehle et al. 1993, 1995), Fra-2 (see above; Baler and Klein 1995), and the β_1-adrenergic receptor, despite a dramatic reduction in NE binding sites.

   In early postnatal life when the pineal parenchyma is not yet or only incompletely innervated by sympathetic nerve fibers (see Auerbach 1982) the amount of β_1-adrenergic receptor mRNA in the rat pineal gland is elevated and not yet regulated. It is only when the sympathetic innervation of the pineal gland gains function at the beginning of the second postnatal week that β_1-adrenergic receptor mRNA levels decrease and become rhythmic. Interestingly, the ontogeny of the rhythm in pineal NAT activity (Ellison et al. 1972; see Klein et al. 1981; Klein 1982), melatonin synthesis (see Klein et al. 1981; Klein 1982), and ICER transcription and inducibility (see below; Stehle et al. 1995; see Stehle 1995) shows a strikingly similar developmental pattern. It can thus be concluded that the maturation of the NE/β_1-adrenergic receptor interaction coincides with a complete maturation of the cAMP-signaling pathway.

2. The activity of TPH, which is readily detectable in the pineal gland and shows elevated levels during the dark period (Ehret et al. 1991), is regulated via β_1-adrenergic receptor activation and can be induced in vitro by cAMP analogs via PKA-dependent protein phosphorylation (Ehret et al. 1991). To date the TPH gene has been cloned from a rat and a rabbit pineal library (Grenett et al. 1987; Darmon et al. 1988). The presence of CRE promoter elements on the 5'-end of the rat pineal TPH gene (Delort et al. 1989) supports its membership in the class of cAMP-inducible genes and suggests, in addition to a regulation of TPH expression at the posttranslational level, a regulation via transient action of CRE-binding transcription factors such as CREB and ICER.

3. Increase and maintenance of NAT depends on transcriptional, translational, and posttranslational mechanisms regulated by elevated intrapineal cAMP levels (Klein 1985; Borjigin et al. 1995; Stehle 1995; Roseboom et al. 1996; Klein et al. 1996). It has therefore long been believed that NAT inducibility rests upon a CRE promoter. Indeed, the recent cloning of the rat NAT gene and the characterization of the gene promoter have revealed that the approx. 100-fold nocturnal increase in NAT enzyme activity is directed transcriptionally by a more than 150-fold increase in NAT mRNA at nighttime (Fig. 27). A nearly perfect CRE (TGACGCCA) has been detected by sequence analysis of the 5'-untranslated region of the NAT gene and characterized by using partial promotor constructs to be sufficient for conferring cAMP responsiveness. However, full NAT induction requires the additionally detected inverted CCAAT box in the NAT promoter. CCAAT boxes are found in many promotors and appear to control gene expression by a positive influence on transcription factor binding capacities.

The importance of transcriptional events for induction of NAT activity varies remarkably on a species-to-species basis. In the sheep pineal organ the level of NAT mRNA varies between day and night only by a factor of 1.5 (Coon et al. 1995; for the chicken, see Bernard et al. 1997a,b), whereas the activity of the enzyme is increased sevenfold in the night over that in the day. Thus mechanisms regulating the induction of the NAT activity seem to be predominantly transcriptional in the rat, but predominantly posttranscriptional in sheep.

In the rat the nocturnal increase in NAT mRNA levels has been found to be regulated by the same mechanisms that account for the induction of the NAT activity: the SCN-driven increase in intrapineal cAMP levels induces the rhythmic NAT transcription. However, the rise in the detectable intrapineal amount of NE (Drijfhout et al. 1996a) and the subsequent increase in NAT mRNA at the beginning of the night (Borjigin et al. 1995; Roseboom et al. 1996) are at least 1 h apart. To explain the time gap between adrenergic stimulation and activation of transcription in the

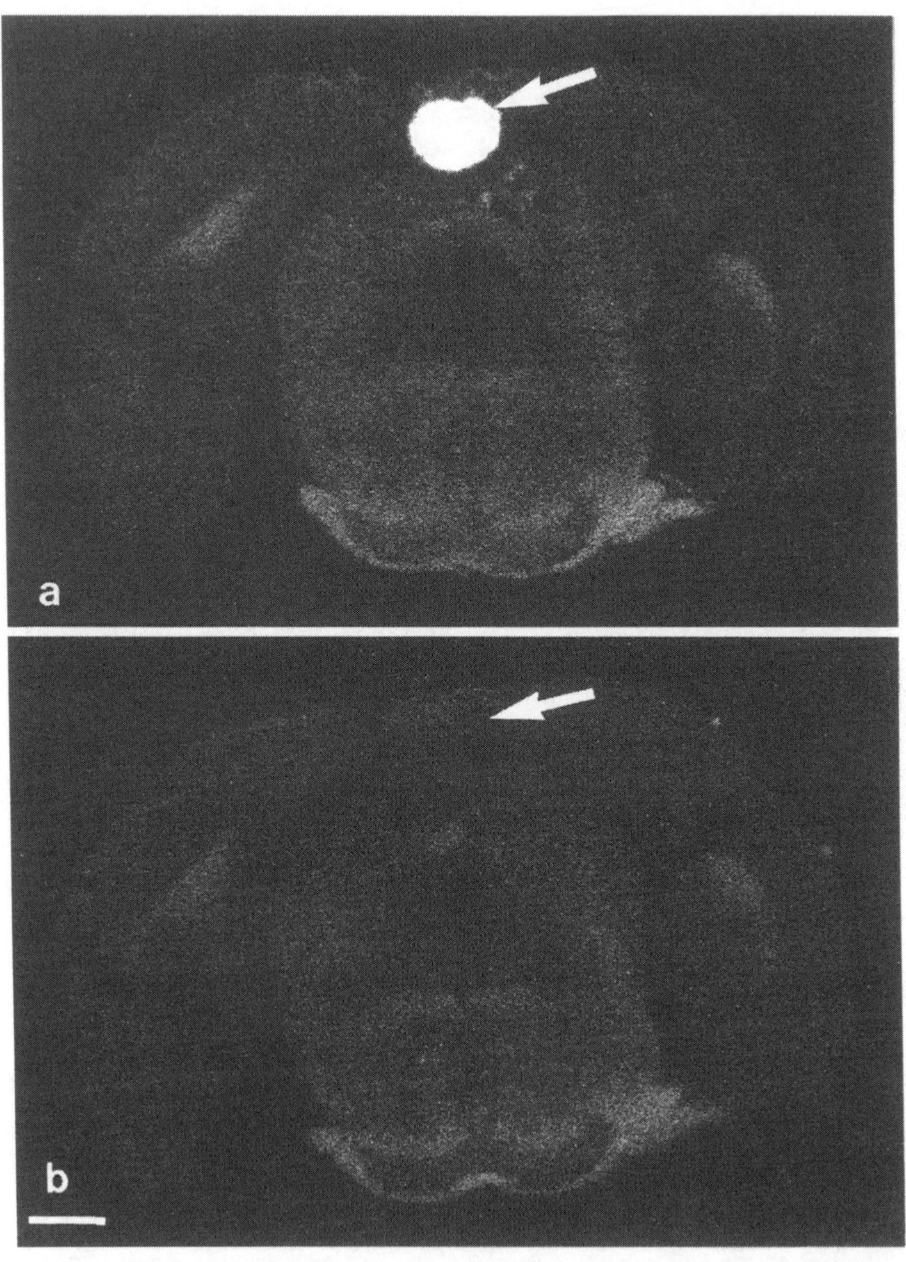

Fig. 27a,b. Autoradiographic demonstration of mRNA for the serotonin-*N*-acetyltransferase in coronal sections of the rat brain by means of in situ hybridization. **a** Strong signal in the pineal organ (*arrow*) of an animal killed at night (7h after lights off). **b** Virtually no signal is present in the pineal (*arrow*) of an animal killed during the day (6h after lights on). ×11 (B. Neeb, C. von Gall, J. Stehle, and H.-W. Korf, unpublished)

rat pineal gland, the cAMP-dependent phosphorylation of a (nuclear) protein was postulated as early as 20 years ago (Winters et al. 1977). Today we know that the phosphorylation of CREB induced by transsynaptic cAMP-signaling bridges this time gap (see above). Compared to other systems (glycogen metabolism in muscle, Cohen 1989) the time required for CREB phosphorylation appears rather prolonged in the pineal: 30 min versus 5–8 s. The different time courses can be explained by the time required for translocation of the catalytic subunit of the PKA into the nucleus (Hagiwara et al. 1993).

The human NAT gene, located on chromosome 17q25 (Coon et al. 1996), is organized into four exons. This structural feature also seems to be present in the rodent gene. Only a single mRNA transcript has been be detected for all NAT genes so far characterized (human, sheep, chicken, rat; Klein et al. 1996), showing a very high homology within the coding region (80% identity). Sequence comparison with acetyltransferases cloned from different tissues so far indicate only few conserved motifs within the pineal NAT, explaining the difficulties in cloning this enzyme by conservative polymerase chain reaction (PCR) approaches with degenerated primers.

Several putative phosphorylation sites are present on the NAT protein, which are likely to be involved in posttranslational modification. For example, light stimuli at night rapidly decrease the NAT enzymatic activity without altering the NAT mRNA levels (Roseboom et al. 1996; Bernard et al. 1997a,b).

4. The activity of HIOMT, the final enzyme of the melatonin synthesis, remains fairly constant throughout a light-dark cycle. Mechanisms that control HIOMT expression are not clearly defined (see Klein et al. 1981). So far HIOMT clones have been successfully rescued from bovine (Ishida et al. 1987; Donohue et al. 1992), chicken (Voisin et al. 1988), and human (Donohue et al. 1993) cDNA libraries. There exists experimental evidence for a role of the maturing pineal sympathetic innervation in the postnatal onset and diurnal regulation of HIOMT activity (Sugden and Klein 1983). Interestingly, the HIOMT gene expressed in the rat pineal organ cannot be activated until the end of the first week after birth (Sugden and Klein 1983) despite the fact that the cAMP-signaling pathway is already inducible at that age by exogenous stimuli (Yuwiler et al. 1977; Stehle et al. 1995). Structural analysis of the human HIOMT gene reveals two distinct 5' ends with different promoters (A and B). Promoter A is exclusively expressed in the retina. With the help of an enhancer, promoter B seems to drive high expression of HIOMT in the pineal. This promoter contains, in addition to an AP-1/TRE site, a possible CRE site (Rodriguez et al. 1995). Thus the transcription of pineal HIOMT may be regulated by the cAMP-signaling pathway and its transcription factors. Recently a transcriptional control of HIOMT has been shown in the rat pineal gland, with a twofold increase in HIOMT-mRNA levels during the dark period (Gauer and Craft 1996).

5. The transcription factor ICER is inducible upon NE activation of the pineal cAMP-signaling pathway. ICER inducibility rests upon the presence of four CREs in the 5'-untranslated part of the transcript (Molina et al. 1993). ICER has been cloned by reverse transcriptase PCR and the cDNA rescued from a nighttime pineal library encodes for a very potent repressor of cAMP-inducible genes (Stehle et al. 1993). ICER is generated as a splicing product of the CREM (CRE-binding protein modulator) gene by use of an internal promoter located downstream of the promoter which generates all other CREM isoforms known so far (Stehle et al. 1993; Molina

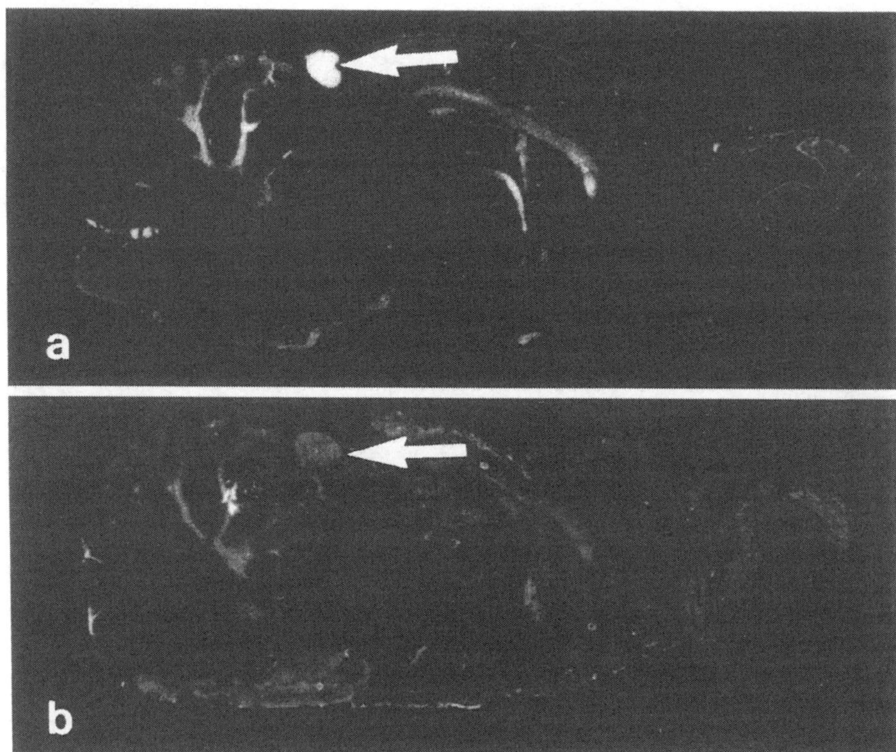

Fig. 28a,b. Autoradiographic demonstration of mRNA for inducible cyclic AMP early repressor (ICER) in midsagittal sections of the rat brain by means of in situ hybridization with a specific antisense riboprobe (see Stehle et al. 1993). **a** Strong signal in the pineal (*arrow*) of an animal killed at night (7h after lights off). **b** Very low signal in the pineal (*arrow*) of an animal killed during the day (5h after lights on). ×3.5

et al. 1993). Due to the use of the ICER-specific internal promoter this CREM transcript is a remarkably small transcription factor essentially comprising the DNA binding domain only. ICER shows a superior affinity to CREs and is a far more potent inhibitor of cAMP-inducible genes than its CREM siblings (CREM α, β, γ; Stehle et al. 1993) when acting as a homodimer. In vivo investigations on the regulation of ICER in the rat pineal gland have revealed a remarkable rhythm in ICER transcription, with elevated levels during the second half of the night which is driven by the SCN and induced by NE (Fig. 28).

The following features are of specific interest with respect to the role of ICER in the pineal gland and the correlation between its expression and the profile of melatonin biosynthesis: (i) The cDNA cloning revealed the absence of the amino-terminal phosphorylation domain in the ICER transcript (P-box; Stehle et al. 1993), although the hallmark of the CREM gene, generation of modular structured isoforms by differential splicing (Foulkes and Sassone-Corsi 1992), is preserved. This indicates that in the pineal gland solely the level of ICER expression determines the activity of this transcriptional repressor. (ii) ICER mRNA is elevated 4–6 h before the end of the dark period and shortly precedes the peak of ICER protein (Stehle et al. 1993). This pattern of expression provides at least a temporal link to the endogenous downregulation of melatonin synthesis (Takahashi 1994; Stehle 1995; Klein et al. 1996). (iii) In vitro and in vivo experiments performed at different times of the day (Stehle et al. 1993) demonstrate this transcriptional repressor to be uninducible at times when ICER is expressed. Thus ICER shows autoregulative features and inhibits its own transcription at the beginning of the light period (Stehle et al. 1993). This grants ICER the properties of a clock protein. Since the mammalian pineal gland does not harbor an endogenous oscillator, the presumed clock function of ICER may be a phylogenetic remnant. Although in rat SCN ICER mRNA does not fluctuate over the 24-h cycle (Stehle et al. 1995, 1996), it may do so in the avian pineal gland that contains an endogenous clock. (iv) In the rat there is a coincidental maturation of CREM inducibility and the endogenous day-night switch in ICER transcription during the early postnatal development (Stehle et al. 1995). As soon as the sympathetic innervation gains function at the beginning of the second postnatal week, melatonin synthesis starts oscillating as does ICER. (v) The level of the catalytic subunit of PKA in rat pineal rises in parallel to ICER expression and melatonin rhythm (Stehle et al. 1995). This indicates an overall maturation of the cAMP-signaling pathway that is retarded in the early ontogenetic phase by an unknown transient blockade, eventually released when pups gain independence from maternal cues. It is tempting to speculate that, despite its very low expression, ICER protein level is sufficient for this block in early postnatal life. This speculation is based on the superior affinity of ICER to CREs and its striking potency in repressing cAMP-inducible genes (Stehle et al. 1993). (vi) ICER is able to heterodimerize with other CREM isoforms and CREB. By heterodimerization ICER may impose its superior affinity to CREs on other transcription factors, but it may also scavenge potentially activating CRE-binding proteins.

The NAT mRNA level is higher in mice bearing a null mutation in the CREM locus than in control animals (Foulkes et al. 1996a). In the rat pineal gland pCREB levels are not yet down to daytime levels when melatonin synthesis decreases endogenously, stressing the function of ICER as a transcriptional downregulator of cAMP-inducible genes. Our current results indicate a dual impact of ICER on the regulation of rat melatonin synthesis: (a) ICER tonically inhibits NAT transcription, with attenuation modulated on the basis of the photoperiodic history (Foulkes et al. 1996b): short photoperiods induce a prolonged and increased nocturnal upregulation in ICER expression as compared to long photoperiods. (b) pCREB and ICER act as competitors for the recently characterized NAT CRE (Foulkes et al. 1996a) at the beginning and the end of the night. Thus a shift in the balance between the two transcription factors leads to an elevated impact of either pCREB or ICER (Fig. 29). Interactions between pCREB and ICER, such as competition for CREs and heterodimerization, may not only shape circadian melatonin synthesis but also provide the basis for dynamic changes required for seasonal adaptation.

Acute alterations in the lighting regimen and progressive waxing and wanning of the dark phase require mechanisms within the pineal cAMP-signaling pathway for both short-term plasticity and long-term adaptation. There is increasing evidence that *cis*-acting transcription factors, such as CREB, FRA-2, and ICER, which are induced via β_1-adrenergic receptor activation, determine and shape the rhythmic melatonin synthesis. The importance of the cAMP-signaling pathway can be anticipated by the

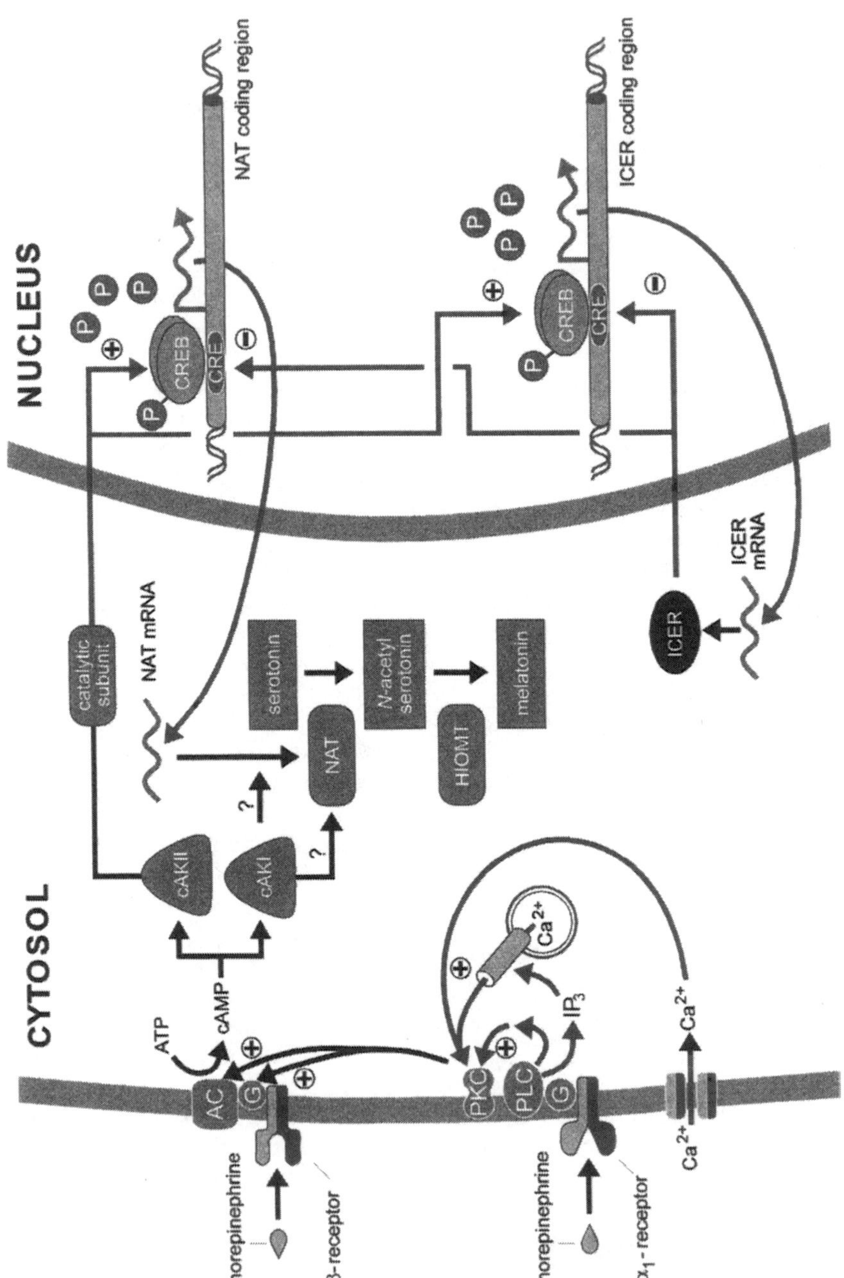

Fig. 29. Schematic diagram of the action of second and third messengers in rat pinealocytes. *AC*, Adenylate cyclase; *G*, GTP-binding protein; *PKC*, protein kinase C; *PLC*, phospholipase C; *IP₃*, inositoltrisphosphate; *cAKI, II*, cyclic AMP-dependent protein kinase types I and II; *NAT*, serotonin (arylalkylamine) *N*-acetyltransferase; *HIOMT*, hydroxyindole-*O*-methyltransferase; *CRE*, cyclic AMP response element; *CREB*, CRE binding protein; *ICER*, inducible cyclic AMP early repressor; *P*, phosphate groups. (Modified and extended after Stehle 1995)

70

dependence of both the circadian rhythm in CREB phosphorylation (J.H. Stehle, E. Maronde, and H.-W. Korf, unpublished) and the circadian profiles in CRE-mediated upregulation of ICER and NAT transcription on the activation of this pathway.

6 Action, Targets, and Receptors of Melatonin

Melatonin is the hormone providing animals and humans with information on the length of the night and thereby also on the length of the day. In amphibians it elicits the primary chromatic response, i.e., blanching of the skin in darkness, by causing pigment aggregation in dermal melanophores (Rollag 1988; Fig. 5). This well-known action was used as a bioassay when Lerner and colleagues isolated melatonin in the late 1950s, and dermal melanophores of the frog *Xenopus laevis* served as the source for the first molecular biological identification of the melatonin receptor (Ebisawa et al. 1994). In birds melatonin is necessary for maintaining normal circadian function and imposes periodicity on structures that ultimately control overt circadian rhythms (Cassone 1990). In mammals melatonin can entrain circadian rhythms and acts in concert with light to keep the circadian rhythm in phase with prevailing environmental conditions (Arendt 1995; Lewy et al. 1992). Thus melatonin is useful to treat jet lag and some circadian sleeping disorders of humans (Arendt 1995). Moreover, melatonin is involved in regulating reproduction in seasonally breeding mammals (Reiter 1991).

Melatonin binding sites have been identified in many vertebrate species by radioreceptor assay and quantitative receptor autoradiography using the specific ligand ^{125}I-labeled melatonin (Vanecek et al. 1987; Vanecek 1988; Dubocovich 1995; Weaver et al. 1989; Morgan et al. 1994). This ligand has been shown to have an even higher biological activity than melatonin itself and thus proved to be an extremely helpful tool in delineating the targets for the pineal hormone and elucidating intracellular signaling pathways affected by melatonin (Morgan et al. 1994).

With this approach high-affinity binding sites for melatonin have been found in distinct nuclei of the brain and some nonneuronal peripheral tissues (Dubocovich 1995). The suprachiasmatic nucleus, which is of particular importance for the organization and functional integrity of the mammalian photoneuroendocrine system, has been shown to contain melatonin binding sites in almost all mammalian species investigated (Fig. 30) including humans (Reppert et al. 1988; Weaver et al. 1993). Melatonin can affect the electrical activity of SCN neurons in vitro (Stehle et al. 1989) and phase-shifts the rhythm of electrical activity when applied in a time window around the light/dark transition (McArthur et al. 1991). More recently melatonin has been found to inhibit the phosphorylation of the transcription factor CREB which was elicited in rat SCN neurons upon stimulation with the neuropeptide PACAP at circadian time 10 (i.e., around the light/dark transition; Kopp et al. 1997).

These findings support the notion that melatonin acts directly through SCN receptors to entrain circadian rhythms. The phase-triggering capacity of melatonin is, however, weak compared to that of light. Functionally more important is the action of melatonin on SCN activity during the prenatal and the early postnatal life, when the

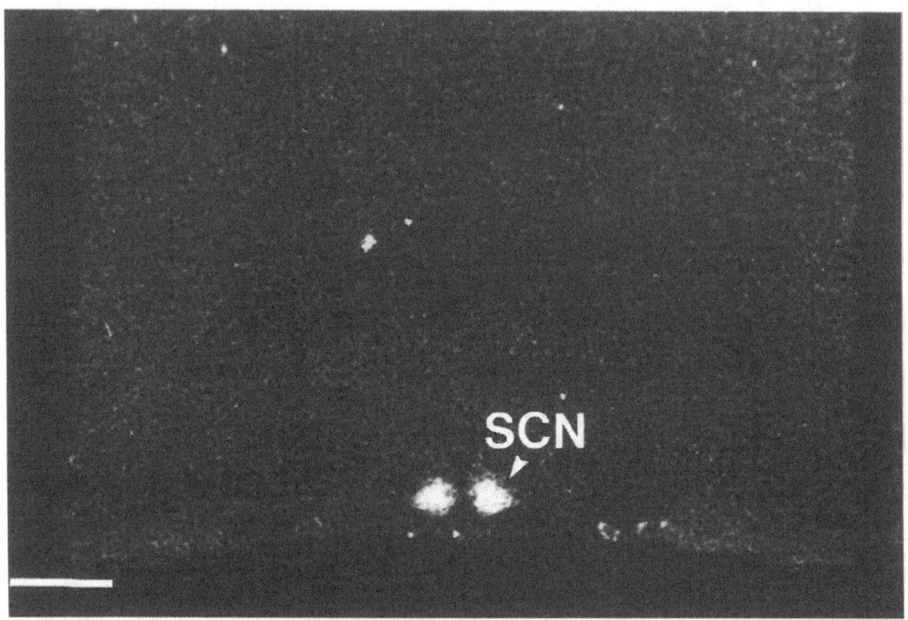

Fig. 30. Autoradiographic demonstration of melatonin binding sites in the rat suprachiasmatic nuclei (*SCN, arrowhead*) using ^{125}I-labeled melatonin as ligand. Coronal section. ×12 (C. von Gall, B. Neeb, H.-W. Korf, and J. Stehle, unpublished)

retinohypothalamic connections are not yet fully established (see Klein et al. 1991). Melatonin derived from the mother readily passes the placental barrier and after birth is taken up with the milk to serve as a communicator of ambient lighting conditions until the the pup's retinohypothalamic tract has gained full functional integrity. Interestingly, melatonin binding sites are even present in the mouse strain C57BL (Siuciak et al. 1990), which is rendered melatonin-deficient by virtue of a genetic defect (Ebihara et al. 1986). These binding sites appear biologically functional since, as in rats (see above), melatonin is found to suppress the PACAP-induced phosphorylation of CREB in the SCN of C57BL mice (C. von Gall, H.-W. Korf and J.H. Stehle, unpublished observations; Fig. 31).

In all seasonally breeding mammals the hypophysial pars tuberalis consistently shows the highest density of melatonin binding sites (Weaver et al. 1991; Morgan et al. 1994). This speaks in favor of the pars tuberalis as the main target site through which melatonin influences the hypothalamic/hypophysial/gonadal axis to determine the timed window(s) for seasonal reproduction. As a mirror of the night length and thereby the daylength, the nocturnal melatonin surge is responsible for the dramatic photoperiodic effects on reproductive status, body weight, coat color and behavior (see Arendt 1995). The fact that melatonin receptors are detected only inconsistently in the pars tuberalis of humans (Weaver et al. 1993) highlights that melatonin's action on neuroendocrine events may differ fundamentally between humans and seasonally breeding species.

For the high-affinity melatonin binding sites an equilibrium dissociation constant (K_D) of less than 100 pM has been determined (Carlson et al. 1989; see Weaver et al.

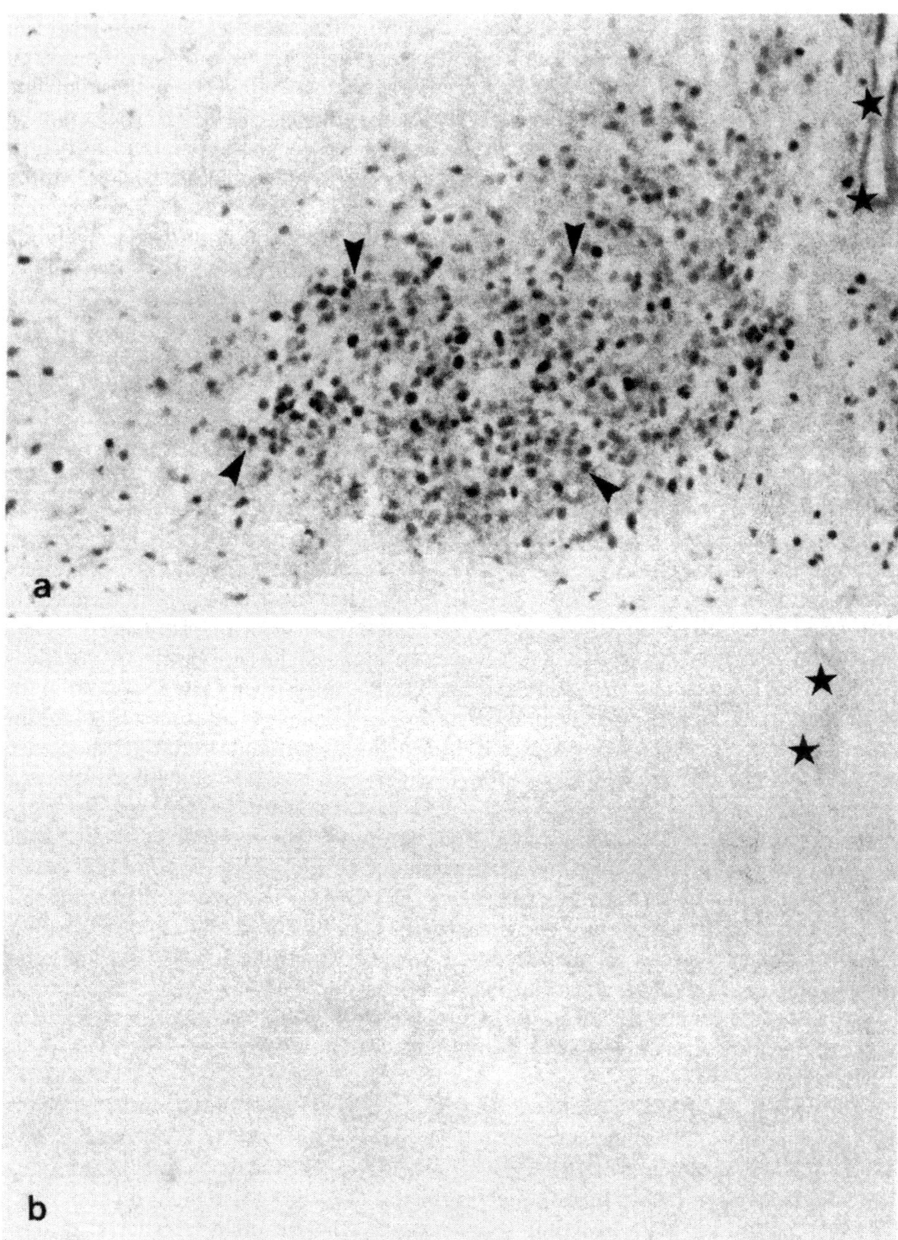

Fig. 31a,b. Effect of melatonin on phosphorylation of CREB in the mouse (C57BL) suprachiasmatic nucleus (SCN). **a** Stimulation of brain slices with 100 n*M* PACAP at circadian time (CT) 10 induced pCREB immunoreactivity in the nuclei of numerous SCN cells (*arrowheads*). **b** Pretreatment of the slices with melatonin (1 n*M*) abolishes this response to PACAP. *Star*, Third ventricle. ×175 (Courtesy C. von Gall; see also Kopp et al. 1997)

1991; Morgan et al. 1994). The affinity of these binding sites to melatonin has been shown to be sensitive to guanine nucleotides, and whenever investigated it has been demonstrated that the activation of this receptor leads to an inhibition of the adenylate cyclase with a subsequent decrease in elevated intracellular cAMP levels. The complete elucidation of the biochemical characteristics and the molecular understanding of how melatonin acts at high-affinity binding sites was hampered for long because of the difficulties in cloning the melatonin receptor(s). This was achieved for the first time when a cDNA was constructed from mRNA extracted from an immortalized cell line of *Xenopus* dermal melanophores (a tissue with an approximately 100 times higher receptor density than the mammalian targets for melatonin) and used for expression cloning (Ebisawa et al. 1994).

Using PCR with degenerated primers based on the frog sequence for the high-affinity melatonin receptor, mammalian melatonin receptors have been cloned from several species (sheep, human: Reppert et al. 1994, 1995a; chick: Reppert et al. 1995b; mouse: Roca et al. 1996). These receptor clones show a 60% identity to the frog receptor clone and are termed Mel_{1a}. Mel_{1a} receptors are expressed in the SCN and the pars tuberalis and are thought to mediate the reproductive and circadian responses of melatonin (Reppert et al. 1994). However, recent experiments showed that transgenic mice in which this receptor type was knocked out were still able to respond to treatment with melatonin with a phase-shift (Liu et al. 1997). Thus the precise type of receptor(s) mediating the circadian effects of melatonin remain(s) to be be determined. A second melatonin receptor, designated Mel_{1b}, with an apparent 60% identity to the Mel_{1a} receptor, has been cloned in human and rat. This receptor is very low abundant in the retina and detectable in brain only by reverse transcriptase PCR; it may mediate the reported physiological actions of melatonin in the mammalian retina (Dubocovich 1995). Sequence comparison has shown that the frog receptor clone encodes for a melatonin receptor subtype, which is distinctly different from the Mel_{1a} and the Mel_{1b} receptors and has been designated as Mel_{1c} receptor. This receptor has so far been detected only in nonmammalian vertebrates such as zebra fish (*Brachidanio rerio*), frog (*Xenopus laevis*), and chicken (Reppert et al. 1995a,b). It is interesting to note that the phylogenetic divergence of the melatonin receptor subfamily within G protein coupled receptors seems to have occurred very early in evolution. For example this unique cluster of receptors shows sequence differences in otherwise highly conserved regions of G protein coupled receptors, i.e., the DRY protein sequence motif downstream the third transmembrane domain is substituted by NRY (Reppert et al. 1995a,b).

All three high-affinity melatonin receptor subtypes, Mel_{1a}, Mel_{1b} and the Mel_{1c}, however, have a very similar gene structure: the receptor protein is encoded by two exons that are spaced apart by a large (>8 kb) intron. The chromosomal location has been delineated for the two human receptors (Mel_{1a} on 4q35.1 and Mel_{1b} on 11q21–22) and for the mouse Mel_{1a} receptor (chromosome 8). The pharmacological characterization in heterologous expression systems has confirmed on the molecular level earlier data (White et al. 1987) that all types of melatonin receptor inhibit forskolin-induced cAMP accumulation by a pertussis toxin-sensitive G protein (Reppert et al. 1995a,b).

In rat gonadotrophs which bear melatonin receptors during early postnatal life (Vanecek 1988), and in which melatonin is able to reverse the stimulatory effects of LHRH on LH release (Martin and Klein 1976) melatonin has been shown to inhibit the

LHRH-induced accumulation of diacylglycerol and release of arachidonic acid, to block Ca^{2+} influx, to cause membrane hyperpolarization, and to repolarize the depolarizing effects of LHRH. These effects are mediated via a pertussis toxin sensitive G protein (Vanecek and Vollrath 1989, 1990; Vanecek and Klein 1992a,b).

Investigations of the signal transduction pathway affected by the human Mel_{1a} receptor expressed in NIH 3T3 cells reveal another impact of melatonin: although melatonin alone does not stimulate the Ca^{2+}-dependent phosphoinositide phospholipid hydrolysis or arachidonic acid release, the hormone can potentiate these effects on cAMP-independent signal transduction pathways elicited by calcium (Godson and Reppert 1997). Also this route is sensitive to pertussis toxin. It can be envisaged that the trimeric inhibitory G protein coupled to the Mel_{1a} receptor directs, on the one hand, the inhibition of cAMP levels via the $G_i\alpha$-subunit and, on the other, activates via the $G_i\beta\gamma$-subunits the phospholipase C in the presence of elevated intracellular Ca^{2+} levels. Thus intracellular second messenger crosstalk may also exist for signaling events induced by melatonin interacting with its receptor subtypes (Godson and Reppert 1997).

7 Summary

The vertebrate pineal organ rhythmically synthesizes and secretes melatonin during nighttime and forms an essential component of the photoneuroendocrine system which allows humans and animals to measure and keep the time. Regulation of the melatonin biosynthesis depends on signals from photoreceptors perceiving and transmitting environmental light stimuli and endogenous oscillators generating a circadian rhythm which is independent from any environmental time cue (zeitgeber). In nonmammalian species the photoreceptors responsible for regulating melatonin biosynthesis reside within the pineal organ itself. In several nonmammalian species (e.g., lamprey, zebra fish, house sparrow, chicken) the pineal organ is also capable of generating circadian rhythms and thus serves all key functions of the photoneuroendocrine system: photoreception, endogenous rhythm generation, and production of neurohormones. These may even be accomplished by a single "photoneuroendocrine" cell.

In mammals the pineal organ has lost both the direct light sensitivity and the capacity of generating circadian rhythms, and melatonin biosynthesis is regulated by retinal photoreceptors and a circadian oscillator located in the suprachiasmatic nucleus of the hypothalamus. Due to this spatial separation the photoneuroendocrine system of mammals comprises neuronal and neuroendocrine pathways which interconnect its components. The neuronal pathways involve circuits of both the central and the peripheral nervous systems, and as an important final link noradrenergic sympathetic nerve fibers. The suprachiasmatic nucleus appears as a major target of melatonin in mammals. The pineal hormone may thus be involved in a feedback loop of the mammalian photoneuroendocrine system.

The present comparative contribution considers, after a short survey of classical findings on the phylogenetic development and the gross anatomy of the pineal complex, cytoevolutionary and cell biological aspects of the various types of pinealocytes as well as the afferent and efferent innervation of the pineal organ (pinealofugal and pinealopetal neuronal pathways). Moreover, emphasis is placed on receptor mechanisms, second messenger systems (Ca^{2+} and cyclic AMP), transcription factors (e.g, CREB and ICER), and their roles for regulation of melatonin biosynthesis. Finally, the action, targets, and receptors of melatonin are dealt with. The synoptic approach of this contribution, which combines anatomical and ultrastructural findings with cell and molecular biological results, confirms the functional significance of the melatonin-synthesizing pineal organ as an important component of the photoneuroendocrine system and stresses the importance of this organ as a model to study signal transduction mechanisms both in photoreceptors and in neuroendocrine cells.

References

Abreu P, Sugden D (1990) Characterization of binding sites for (^{3}H)-DTG, a selective sigma receptors ligand, in the sheep pineal gland. Biochem Biophys Res Commun 171:875–881

Aloyo VJ (1991) Preproenkephalin A gene expression in rat pineal. Neuroendocrinology 54:594–598

Aloyo VJ, Walker RF (1987) Noradrenergic stimulation of serotonin release from rat pineal glands in vitro. J Endocrinol 114:3–9

Aloyo VJ, Walker RF (1988) Alpha-adrenergic control of serotonin release from rat pineal glands. Neuroendocrinology 48:61–66

Antoch MP, Song EJ, Chang AM, Vitaterna MH, Zhao Y, Wilsbacher LD, Sangoram AM, King DP, Pinto LH, Takahashi JS (1997) Functional identification of the mouse circadian Clock gene by transgenic BAC rescue. Cell 89:655–667

Antonow A (1925) Zur Frage von dem Bau der Glandula pinealis. Anat Anz 60:21–31

Araki M (1992) Cellular mechanism for norepinephrine suppression of pineal photoreceptor-like cell differentiation in rat pineal cultures. Dev Biol 149:440–447

Araki M, Tokunaga F (1990) Norepinephrine suppresses both photoreceptor and neuron-like properties expressed by cultured rat pineal glands. Cell Diff Dev 31:129–135

Araki M, Watanabe K, Tokunaga F, Nonaka T (1988) Phenotypic expression of photoreceptor and endocrine cell properties by cultured pineal cells of the newborn rat. Cell Diff Dev 25:155–164

Arendt J (1995) Melatonin and the mammalian pineal gland. Chapman and Hall, London

Arstila AU (1967) Electron microscopic studies on the structure and histochemistry of the pineal gland of the rat. Neuroendocrinology 2:1–101

Auerbach DA (1982) β-Adrenergic receptors during development. In: Klein DC (ed) Melatonin rhythm generating system. Basel, Karger, pp 97–107

Axelrod J (1974) The pineal gland: a neurochemical transducer. Science 184:1341–1348

Axelrod J, Weissbach H (1960) Enzymatic O-methylation of N-acetyl-serotonin to melatonin. Science 131:1312

Axelrod J, Weissbach H (1961) Purification and properties of hydroxyindole-O-methyltransferase. J Biol Chem 236:211–213

Babila T, Schaad NC, Klein DC (1992) Rat pineal G$_s\alpha$, G$_i\alpha$ and G$_o\alpha$: relative abundance and development. Brain Res 572:232–235

Baler R, Klein DC (1995) Circadian expression of transcription factor Fra-2 in the rat pineal gland. J Biol Chem 270:27319–27325

Baler R, Covington S, Klein DC (1997) The rat arylalkylamine N-acetyltransferase gene promoter. J Biol Chem 272:6979–6985

Bargmann W (1943) Die Epiphysis cerebri. In: Von Möllendorff W (ed) Handbuch der mikroskopischen Anatomie des Menschen. Springer, Berlin Heidelberg New York, pp 309–502

Barry J (1979) Immunofluorescence study of the preoptico-terminal LHRH tract in the female squirrel monkey during the esterous cycle. Cell Tissue Res 198:1–13

Bégay V, Bois P, Collin JP, Lenfant J, Falcon J (1994) Calcium and melatonin production in dissociated trout pineal photoreceptor cells in culture. Cell Calcium 16:37–46

Benovic JL, Mayor F, Somers RL, Caron MG, Lefkowitz RJ (1986) Light-dependent phosphorylation of rhodopsin by beta-adrenergic receptor kinase. Nature 321:869–872

Benovic JL, Kühn H, Weyand I, Codina J, Carion MG, Lefkowitz RJ (1987) Functional desensitization of the isolated β-adrenergic receptor by the β-adrenergic receptor kinase: potential role of an analog of the retinal protein arrestin (48-kDa protein). Proc Natl Acad Sci USA 84:8879–8882

Bernard M, Voisin P, Guerlotte J, Collin JP (1991) Molecular and cellular aspects of hydoxyindole-O-methyltransferase expression in the developing chick pineal gland. Dev Brain Res 59:75–81

Bernard M, Klein DC, Zatz M (1997a) Chick pineal clock regulates serotonin N-acetyltransferase mRNA rhythm in culture. Proc Natl Acad Sci USA 94:304–309

Bernard M, Iuvone PM, Cassone VM, Roseboom PH, Coon SL, Klein DC (1997b) Melatonin synthesis: photic and circadian regulation of serotonin N-acetyltransferase mRNA in the chicken pineal gland and retina. J Neurochem 68:213–224

Binkley S, Reibman JB, Reilly KB (1978) The pineal gland: a biological clock in vitro. Science 202:1198–1201

Björklund A, Owman C West KA (1972) Peripheral sympathetic innervation and serotonin cells in the habenular region of the rat brain. Z Zellforsch 127:570–579

Blank HM, Müller B, Korf HW (1997) Comparative investigations of the neuronal apparatus in the pineal organ and retina of the rainbow trout: immunocytochemical demonstration of neurofilament 200-kDa and neuropeptide Y, and tracing with DiI. Cell Tissue Res 288:417–425

Bönigk W, Müller F, Middendorff R, Weyand I, Kaupp UB (1996) Two alternatively spliced forms of the cGMP-gated channel α-subunit from the cone photoreceptor are expressed in the chick pineal organ. J Neurosci 16:7458–7468

Borjigin J, Wang MM, Snyder SH (1995) Diurnal variation in mRNA encoding serotonin-N-acetyl-transferase in pineal gland. Nature 378:783–785

Bowers CW, Zigmond RE (1982) The influence of the frequency and pattern of sympathetic nerve activity on serotonin-N-acetyl-transferase in the rat pineal gland. J Physiol (Lond) 330:279–296

Bowers CW, Dahm LM, Zigmond RE (1984) The number and distribution of sympathetic neurons that innervate the rat pineal gland. Neuroscience 13:87–96

Buijs RM, Pévet P (1980) Vasopressin- and oxytocin-containing nerve fibers in the pineal gland and subcommissural organ of the rat. Cell Tissue Res 205:11–17

Cahill (1996) Circadian regulation of melatonin production in cultured zebrafish pineal and retina. Brain Res 708:177–181

Cajal RS (1904) Textura del Sistema Nervioso del Hombre y de los Vertebrados, TII, 2da parte. Moya, Madrid

Cantor EH, Greenberg LH, Weiss B (1981) Effect of long-term changes in sympathetic nervous activity on the beta-adrenergic receptor-adenylate cyclase complex of the rat pineal gland. Mol Pharmacol 19:21–26

Cardinali DP, Vacas MI, Rosenstein RE, Etchegoyen GS, Sariento MIK, Solveyra CG, Pereyra EN (1987) Multifactorial control of pineal melatonin synthesis: an analysis through binding studies. Adv Pineal Res 2:51–66

Carlson LL, Weaver DR, Reppert SM (1989) Melatonin signal transduction in hamster brain: inhibition of adenylyl cyclase by pertussis toxin-sensitive G protein. Endocrinology 125:2670–2676

Carter DA (1992) Neurotransmitter-stimulated immediate-early gene responses are organized through differential post-synaptic receptor mechanisms. Mol Brain Res 16:111–118

Carter DA (1993) Noradrenergic regulation of c-jun expression in the rat pineal gland in culture: positive and negative components. Eur J Pharmacol 247:97–100

Carter DA (1994) A daily rhythm of activator protein 1-activity in the rat pineal gland is dependent upon trans-synaptic induction of jun-B. Neuroscience 62:1267–1278

Cassone VM (1990) Effects of melatonin on vertebrate circadian systems. Trends Neurosci 13:457–464

Cassone VM, Menaker M (1983) Sympathetic regulation of chicken pineal rhythms. Brain Res 272:311–317

Cassone VM, Takahashi JS, Blaha CD, Lane RF, Menaker M (1986) Dynamics of noradrenergic input to the chicken pineal gland. Brain Res 384:334–341

Ceña V, Halperin JI, Yeandle S, Klein DC (1991) Norepinephrine stimulates potassium efflux from pinealocytes: evidence for involvement of biochemical "AND" gate operated by calcium and adenosine 3',5'-monophosphate. Endocrinology 128:559–569

Chik CL, Ho AK (1990) Multiple receptor regulation of cyclic nucleotides in rat pinealocytes. Prog Biophys Mol Biol 53:197–203

Chik CL, Ho AK (1995) Pituitary adenylate cyclase-activating polypeptide: control of rat pineal cyclic AMP and melatonin but not cyclic GMP. J Neurochem 64:2111–2117

Chik C, Ho A, Klein DC (1988) Dual receptor regulation of cyclic nucleotides: α_1-adrenergic potentiation of vasoactive intestinal peptide stimulation of pinealocyte adenosine 3',5'-monophosphate. Endocrinology 122:1646–1651

Cohen P (1989) The structure and regulation of protein phosphatases. Annu Rev Biochem 58:453–508

Collin JP (1971) Differentiation and regression of the cells of the sensory line in the epiphysis cerebri. In: Wolstenholme GEW, Knight J (eds) The pineal gland. Churchill-Livingstone, Edinburgh, pp 79–125

Collin JP, Oksche A (1981) Structural and functional relationships in the nonmammalian pineal gland. In: Reiter RJ (ed) The pineal gland, vol 1. Anatomy and biochemistry. CRC, Boca Raton, pp 27–67

Coon SL, Roseboom PH, Baler R, Weller JL, Namboodiri MAA, Koonin EV, Klein DC (1995) Pineal serotonin-N-acetyltransferase: expression cloning and molecular analysis. Science 270:1681–1683

Coon SL, Mazuruk K, Bernard M, Klein DC (1996) The human serotonin N-acetyltransferase (EC 2.3.1.87) gene: structure, chromosomal localization and tissue expressions. Genomics 34:76–84

Coto-Montes A, Masson-Pevet M, Pevet P, Møller M (1994) The presence of opioidergic pinealocytes in the pineal gland of the European hamster (Cricetus cricetus): an immunocytochemical study. Cell Tissue Res 278:483–491

Cozzi B, Mikkelsen JD, Merati D, Capsoni S, Møller M (1989) Vasoactive intestinal peptide (VIP)-like immunoreactive nerve fibers in the pineal gland of the sheep. J Pineal Res 8:41–47

Cozzi B, Mikkelsen JD, Ravault JP, Møller M (1992) Neuropeptide Y (NPY) and C-flanking peptide of NPY in the pineal gland of normal and ganglionectomized sheep. J Comp Neurol 316:238–250

Curran T, Abate C, Cohen DR, MacGregor PF, Rauscher III J, Sonnenberg JL, Connor JA, Morgan JI (1990) Inducible proto-oncogene transcription factors: third messengers in the brain? Cold Spring Harbor Laboratory, Cold Spring Harbor, pp 225–234 (Cold Spring Harbor Symposia on Quantitative Biology, vol LV)

Darmon MC, Guibert B, Leviel V, Ehret M, Maitre M, Mallet J (1988) Sequence of two mRNAs encoding active rat tryptophan hydroxylase. J Neurochem 51:312–316

David GFX, Herbert J (1973) Experimental evidence for a synaptic connection between habenula and pineal ganglion in the ferret. Brain Res 64:327–343

David GFX, Umberkoman B, Kumar K, Anand Kumar TC (1975) Neuroendocrine significance of the pineal. In: Knigge KM, Scott DE, Kobayashi H, Ishii S (eds) Brain-endocrine interaction II. Karger, Basel, pp 365–375

Deguchi T (1979) A circadian oscillator in cultured cells of chicken pineal gland. Nature 282:94–96

Deguchi T (1981) Rhodopsin-like photosensitivity of isolated chicken pineal gland. Nature 290:706–707

Delort J, Dumas JB, Darmon MC, Mallet J (1989) An efficient strategy for cloning 5'extremities of rare transcripts permits isolation of multiple 5'untranslated regions of rat tryptophan hydroxylase mRNA. Nucleic Acid Res 17:6439–6448

Dodt E (1963) Photosensitivity of the pineal organ in the teleost, Salmo irideus Gibbons. Experientia 19:53–61

Dodt E (1973) The parietal eye (pineal and parapineal organs) of lower vertebrates. In: Jung R (ed) Handbook of sensory physiology, vol VIII/3B. Springer, Berlin Heidelberg New York, pp 113–140

Dodt E, Heerd E (1962) Mode of action of pineal nerve fibers in frogs. J Neurophysiol 25:405–429

Dohlman HG, Caron MG, Lefkowitz RJ (1987) Structure and function of the beta-2 adrenergic receptor homology with rhodopsin. Kidney Int 32:1–7

Donohue SJ, Roseboom P, Klein DC (1992) Bovine hydroxyindole-O-methyltransferase. J Biol Chem 267:5184–5185

Donohue SJ, Roseboom PH, Illnerova H, Weller JC, Klein DC (1993) Human hydroxyindole-O-methyltransferase: presence of line-1 fragment in a cDNA clone and pineal mRNA. DNA Cell Biol 12:715–727

Drijfhout WJ, van der Linde AG, Kooi SE, Grol CJ, Westerink BHC (1996a) Norepinephrine release in the rat pineal gland: the input from the biological clock measured by in vivo microdialysis. J Neurochem 66:748–755

Drijfhout WJ, Grol CJ, Westerink BHC (1996b) Parasympathetic inhibition of pineal indole metabolism by prejunctional modulation of noradrenergic release. Eur J Pharmacol 308:117–124

Dryer SE, Henderson D (1991) A cyclic GMP-activated channel in dissociated cells of the chick pineal gland. Nature 353:756–758

D'Souza T, Dryer SE (1994) Intracellular free Ca^{2+} in dissociated cells of the chick pineal gland: regulation by membrane depolarization, second messengers and neuromodulators, and evidence for release of intracellular Ca^{2+} stores. Brain Res 656:85–94

D'Souza T, Dryer SE (1996) A cationic channel regulated by a vertebrate intrinsic circadian oscillator. Nature 382:165–167

Dubocovich ML (1995) Melatonin receptors: are there multiple subtypes? Trends Pharmacol Sci 16:50–56

Ebadi M, Govitrapong P (1986) Orphan transmitters and their receptor sites in the pineal gland. Pineal Res Rev 4:1–54

Ebadi M, Hexum TD, Pfeiffer RF, Govitrapong P (1989) Pineal and retinal peptides and their receptors. Pineal Res Rev 7:1–156

Ebihara S, Marks T, Hudson DJ, Menaker M (1986) Genetic control of melatonin synthesis in the pineal gland of the mouse. Science 231:431–433

Ebisawa T, Karne S, Lerner MR, Reppert SM (1994) Expression cloning of a high affinity melatonin receptor from Xenopus dermal melanophores. Proc Natl Acad Sci USA 91:6133–6137

Ehret M, Pevet P, Maitre M (1991) Tryptophan hydroxylase synthesis is induced by 3',5'cyclic adenosine monophosphate during circadian rhythm in the rat pineal gland. J Neurochem 57:1516–1521

Ekblad E, Edvinsson L, Wahlestedt C, Uddman R, Håkanson R, Sundler F (1984) Neuropeptide Y co-exists and co-operates with noradrenaline in perivascular nerve fibers. Regul Pept 8:225–235

Ekström P (1984) Central neural connections of the pineal organ and retina in the teleost Gasterosteus aculeatus. J Comp Neurol 226:321–336

Ekström P (1987) Photoreceptors and CSF-containing neurons in the pineal organ of a teleost fish have direct axonal connections with the brain: a HRP-electronmicroscopic study. J Neurosci 7:987–995

Ekström P, Korf HW (1985) Pineal neurons projecting to the brain of the rainbow trout, Salmo gairdneri Richardson (teleostei). In-vitro retrograde filling with horseradish peroxidase. Cell Tissue Res 240:693–700

Ekström P, Korf HW (1986) Substance P-like immunoreactive neurons in the photosensory pineal organ of the rainbow trout, Salmo gairdneri Richardson (Teleostei). Cell Tissue Res 246:359–364

Ekström P, Meissl H (1990) Electron microscopic analysis of S-antigen- and serotonin-immunoreactive neural and sensory elements in the photosensory pineal organ of the salmon. J Comp Neurol 292:73–82

Ekström P, Meissl H (1997) The pineal organ of teleost fishes. Rev Fish Biol Fisheries 7:199–284

Ekström P, Vanecek J (1992) Localization of 2-$[^{125}I]$ iodomelatonin binding sites in the brain of the atlantic salmon, Salmo salar L. Neuroendocrinology 55:529–537

Ekström P, Foster RG, Korf HW, Schalken JJ (1987a) Antibodies against retinal photoreceptor-specific proteins reveal axonal projections from the photosensory pineal organ in teleosts. J Comp Neurol 265:25–33

Ekström P, van Veen T, Bruun A, Ehinger B (1987b) GABA-immunoreactive neurons in the photosensory organ of the rainbow trout: two distinct neuronal populations. Cell Tissue Res 250:87–92

Ekström P, Honkanen T, Ebbesson SOE (1988) FRMF-amide like immunoreactive neurons of the nervus terminalis of teleosts innervate both retina and pineal organ. Brain Res 460:68–75

Eldred WD, Nolte J (1981) Multiple classes of photoreceptors and neurons in the frontal organ of Rana pipiens. J Comp Neurol 203:269–296

Eldred WD, Finger TE, Nolte J (1980) Central projections of the frontal organ of Rana pipiens, as demonstrated by the anterograde transport of horseradish peroxidase. Cell Tissue Res 211:215–222

Ellison N, Weller JL, Klein DC (1972) Development of a circadian rhythm in the activity of pineal serotonin N-acetyltransferase. J Neurochem 19:1334–1341

Falck B, Hillarp NA, Thieme G, Torp A (1962) Fluorescence of catecholamines and related compounds with formaldehyde. J Histochem Cytochem 10:348–354

Falcon J (1984) Identification et propriétés des cellules photoneuroendocrines de l'organe pinéal. Thesis, University of Poitiers

Falcón J, Thibault C, Martin C, Brun-Marmillon J, Claustrat B, Collin JP (1991) Regulation of melatonin production by catecholamines and adenosine in a photoreceptive pineal organ. An in vitro study in the pike and in the trout. J Pineal Res 11:123–134

Falcon J, Begay V, Goujon JM, Voisin P, Guerlotte J, Collin JP (1994) Immunocytochemical localization of hydroxyindole-O-methyltransferase in pineal photoreceptor cells of several fish species. J Comp Neurol 341:559–566

Foster RG, Korf HW, Schalken JJ (1987) Immunocytochemical markers revealing retinal and pineal but not hypothalamic photoreceptor systems in the Japanese quail. Cell Tissue Res 248:161–167

Foster RG, Schalken JJ, Timmers AM, De Grip WJ (1989a) A comparison of some photoreceptor characteristics in the pineal and retina. I. The Japanese quail (Coturnix coturnix). J Comp Physiol [A] 165:553–563

Foster RG, Timmers AM, Schalken JJ, de Grip WJ (1989b) A comparison of some photoreceptor characteristics in the pineal and retina. II. The Djungarian hamster (Phodopus sungorus). J Comp Physiol [A] 165:565–572

Foulkes NS, Sassone-Corsi P (1992) More is better: activators and repressors from the same gene. Cell 68:411–414

Foulkes NS, Borjigin J, Snyder SH, Sassone-Corsi P (1996a) Transcriptional control of circadian hormone synthesis via the CREM feedback loop. Proc Natl Acad Sci USA 93:14140–14145

Foulkes NS, Duval G, Sassone-Corsi P (1996b) Adaptive inducibility of CREM as transcriptional memory of circadian rhythms. Nature 381:83–85

Gauer F, Craft CM (1996) Circadian regulation of hydoxyindole-O-methyltransferase mRNA levels in rat pineal and retina. Brain Res 737:99–109

Gern WA, Greenhouse SS (1988) Examination of in vitro melatonin secretion from superfused trout (Salmo gairdneri) pineal organs maintained under dial illumination and continuous darkness. Gen Comp Endocrinol 71:163–174

Gern WA, Greenhouse SS, Nervina JM, Gasser PJ (1992) The rainbow trout pineal organ: an endocrine photometer. In: Ali MA (ed) Rhythms in fishes. Plenum, New York, pp 199–218

Ginty DD, Kornhauser JM, Thompson MA, Bading H, Mayo K, Takahashi JS, Greenberg ME (1993) Regulation of CREB phosphorylation in the suprachiasmatic nucleus by light and a circadian clock. Science 260:238–241

Godson C, Reppert SM (1997) The Mel1a receptor is coupled to parallel signal transduction pathways. Endocrinology 138:397–404

Gonzales GA, Montminy M (1989) Cyclic AMP stimulates somatostatin gene transcription by phosphorylation of CREB at serine 133. Cell 59:675–680

Govitrapong P, Ebadi M (1988) The inhibition of pineal arylalkylamine N-acetyltransferase by glutamic acid and its analogues. Neurochem Int 13:223–230

Govitrapong P, Ebadi M, Murrin LC (1986) Identification of Cl^-/Ca^{2+}-dependent glutamate (quisqualate) binding sites in bovine pineal organ. J Pineal Res 3:223–234

Grennet HE, Ledley FD, Reed LL, Woo SLC (1987) Full-length cDNA for rabbit tryptophan hydroxylase: functional domains and evolution of aromatic acid hydroxylases. Proc Natl Acad Sci USA 84:5530–5534

Guerlotte J, Falcon J, Voisin P, Collin JP (1986) Indoles in the photoreceptor complex of the lamprey pineal complex. Ann Endocrinol 47:62–64

Gwinner (1990) Significance of melatonin and the pineal organ in the control of avian circadian systems. In: Bell BD, Cossee RO, Flux JEC, Heather BD, Hitchmouth RA, Robertson CJR, Williams MJ (eds) Acta Congressus internationalis Ornithologici, vol IV. Ornithological Congress, Trust Board, New Zealand, pp 2022–2029

Hafeez MA, Korf HW, Oksche A (1987) Immunocytochemical and electron microscopic investigations of the pineal organ in adult agamid lizards, Uromastix hardwicki. Cell Tissue Res 250:571–578

Hafeez MA, Naz Y, Korf HW (1995) Immunocytochemical characterisation of retinal photoreceptors and pinealocytes in saurian and ophidian species. J Anat 187:227

Hagiwara M, Alberts A, Brindle P, Meinkoth J, Feramisco J, Deng T, Karin M, Shenolikar S, Montminy M (1992) Transcriptional attenuation following cAMP induction requires PP-1 mediated dephosphorylation of CREB. Cell 70:105–113

Hagiwara M, Brindle P, Harootunian A, Armstrong P, Rivier J, Vale W, Tsien R, Montminy M (1993) Coupling of hormonal stimulation and transcription via the cyclic AMP-responsive factor CREB is rate-limited by the nuclear entry of protein kinase A. Mol Cell Biol 13:4852–4859

Hall JC (1995) Tripping along the trail to the molecular mechanisms of biological clcoks. Trends Neurosci 18:230–240

Harrison N, Zatz M (1989) Voltage-dependent calcium channels regulate melatonin output from cultured chick pineal cells. J Neurosci 9:2462–2467

Hartmann F (1957) Über die Innervation der Epiphysis cerebri einiger Säugetiere. Z Zellforsch 46:416–429

Hausdorff WP, Caron MG, Lefkowitz RJ (1990) Turning off the signal: desensitization of β-adrenergic function. FASEB J 6:2323–2332

Hedlund L, Nalbandov AV (1969) Innervation of the avian pineal body. Am Zool 9:1090

Henderson D, Dryer SE (1992) Voltage- and Ca^{2+}-activated ionic currents in acutely dissociated cells of the chick pineal gland. Brain Res 572:182–189

Hewing M (1978) A liquor contacting area in the pineal recess of the golden hamster (*Mesocricetus auratus*). Anat Embryol (Berl) 153:295–304

Hirunagi K, Ebihara S, Okano T, Takanaka Y, Fukuda Y (1997) Immunoelectron-microscopic investigation of the subcellular localization of pinopsin in the pineal organ of chicken. Cell Tissue Res 289:235–241

Ho AK, Chik CL, Klein DC (1987) Protein kinase C is involved in adrenergic stimulation of pineal cGMP accumulation. J Biol Chem 262:10059–10064

Hollwich F (1979) The influence of ocular light perception on metabolism in man and in animal. In: Schäfer KE (ed) Topics in environmental physiology and medicine. Springer, Berlin Heidelberg New York, pp 1–129

Huang HT, Lin HS (1984) Synaptic junctions between the adrenergic axon varicosity and the pinealocyte in the rat. J Pineal Res 1:281–291

Huang SK, Taugner R (1984) Gap junctions between guineapig pinealocytes. Cell Tissue Res 235:137–141

Huang SK, Klein DC, Korf HW (1992) Immunocytochemical demonstration of rod-opsin, S-antigen, and neuron-specific proteins in the human pineal gland. Cell Tissue Res 267:493–498

Humbert W, Pévet P (1991) Calcium content and concretions of pineal glands of young and old rats. Cell Tissue Res 263:593–596

Ishida I, Obinata M, Deguchi T (1987) Molecular cloning and nucleotide sequence of cDNA encoding hydroxyindole-O-methyltransferase of bovine pineal glands. J Biol Chem 262:2895–2899

Jaim-Etcheverry G, Zieher LM (1971) Ultrastructural cytochemistry and pharmacology of 5-hydroxytryptamine in adrenergic nerve endings. III. Selective increase of norepinephrine in the rat pineal gland consecutive to depletion of neuronal 5-hydroxytryptamine. J Pharmacol Exp Ther 178:42–48

Jaim-Etcheverry G, Zieher LM (1980) Stimulation-depletion of serotonin and noradrenaline from vesicles of sympathetic nerves in the pineal gland of the rat. Cell Tissue Res 207:13–20

Jaim-Etcheverry G, Zieher LM (1983) Ultrastructural evidence for monoamine uptake by vesicles of pineal sympathetic nerves immediately after their stimulation. Cell Tissue Res 233:463–470

Jansen KLR, Dragunow M, Faull RLM (1990) Sigma receptors are highly concentrated in the rat pineal gland. Brain Res 507:158–160

Kaku K, Tsuchiya M, Matsuda M, Inoue Y, Kaneko T, Yanaihara N (1985) Light and agonist alter vasoactive intestinal peptide binding and intracellular accumulation of adenosine 3',5'-monophosphate in the rat pineal gland. Endocrinology 117:2371–2375

Kaku K, Tsuchiya M, Tanizawa Y, Okuya S, Indoue Y, Kaneko T, Yanaihara N (1986) Circadian cycles in VIP content and VIP stimulation of cyclic AMP accumulation in the rat pineal gland. Peptides 7:193–195

Kaku K, Harada Y, Okubo M, Yaga K, Yanaihara N, Kaneko T (1992) Helodermin stimulates intracellular accumulation of cyclic AMP and N-acetyltransferase activity in rat pineal gland. Biomed Res 13:191–195

Kalsow CM, Greenhouse SS, Gern W, Adamus G, Hargrave PA, Lang LS, Donoso LA (1991) Photoreceptor cell specific proteins of snake pineal. J Pineal Res 11:49–56

Kaneko T, Cheng PY, Oka H, Oda T, Yanaihara N, Yanaihara C (1980) Vasoactive intestinal polypeptide stimulates adenylate cyclase and serotonin N-acetyltransferase activities in rat pineal gland in vitro. Biomed Res 1:84–87

Kappers JA (1960) The development, topographical relations and innervation of the epiphysis cerebri in the albino rat. Z Zellforsch 52:163–215

Kappers JA (1965) Survey of the innervation of the epiphysis cerebri and the accessory pineal organs of vertebrates. Prog Brain Res 10:87–153

Kappers JA (1969) The mammalian pineal organ. J Neuro Visc Relat 9:140–184

Kenny GCT (1961) The "nervus conarii" of the monkey. An experimental study. J Neuropathol Exp Neurol 20:563–570

King DP, Zhao Y, Sangoram AM, Wilsbacher LD, Tanaka M, Antoch MP, Steeves TDL, Vitaterna MH, Kornhauser JM, Lowrey PL, Turek FW, Takahashi JS (1997) Positional cloning of the mouse circadian Clock gene. Cell 89:641–654

Klein DC (1982) Melatonin rhythm generating system. Developmental aspects. Karger, Basel, pp 1–249

Klein DC (1985) Photoneural regulation of the mammalian pineal gland. In: Evered D, Clark S (eds) Photoperiodism, melatonin and the pineal gland. Pitman, London, pp 38–56

Klein DC, Weller JL (1970) Indole metabolism in the pineal gland: a circadian rhythm in N-acetyltransferase. Science 169:1093–1095

Klein DC, Auerbach DA, Namboodiri MAA, Wheler GHT (1981) Indole metabolism in the mammalian pineal gland. In: Reiter RJ (ed) The pineal gland, vol I. Anatomy and biochemistry. CRC, Boca Raton, pp 199–227

Klein DC, Moore RY, Reppert SM (1991) Suprachiasmatic nucleus: the mind's clock. Oxford Press, New York

Klein DC, Schaad NL, Namboordiri MAA, Weller JL (1992) Control of N-acetyltransferase. Biochem Soc Trans 20:299–304

Klein DC, Roseboom PH, Coon SL (1996) New light is shining on the melatonin rhythm enzyme. The first postcloning review. Trends Endocrinol Metab 7:106–112

Koistinaho J, Yang G (1990) Induction of c-fos protein-like immunoreactivity in the rat and hamster pineal gland after the onset of darkness. Histochemistry 95:73–76

Kolmer W, Löwy R (1922) Beiträge zur Physiologie der Zirbeldrüse. Pflugers Arch Ges Physiol 196:1–14

Kopp M, Meissl H, Korf HW (1997) The pituitary adenylate cyclase-activating polypeptide-induced phoshorylation of the transcription factor CREB (cAMP response element binding protein) in the rat suprachiasmatic nucleus is inhibited by melatonin. Neurosci Lett 227:145–148

Korf B, Rollag MD, Korf HW (1989) Ontogenetic development of S-antigen- and rod-opsin immunoreactions in retinal and pineal photoreceptors of Xenopus laevis in relation to the onset of melatonin-dependent color-change mechanisms. Cell Tissue Res 258:319–329

Korf HW (1974) Acetylcholinesterase-positive neurons in the pineal and parapineal organs of the rainbow trout, Salmo gairdneri (with special reference to the pineal tract). Cell Tissue Res 154:475–489

Korf HW (1976) Histological, histochemical and electron microscopic studies on the nervous apparatus of the pineal organ of the tiger salamander, Ambystoma tigrinum. Cell Tissue Res 174:475–497

Korf HW (1994) The pineal organ as a component of the biological clock. Ann NY Acad Sci 719:13–42

Korf HW (1996) Innervation of the pineal gland. In: Burnstock G (ed) Series on the autonomic nervous system, vol 10. Autonomic-endocrine interactions [Unsicker K (ed)]. Harwood, Amsterdam, pp 129–180

Korf HW, Ekström P (1987) Photoreceptor differentiation and neuronal organization of the pineal organ. In: Trentini GP, Gaetani C de, Pévet P (eds) Fundamentals and clinics in pineal research. Raven, NewYork, pp 35–47

Korf HW, Møller M (1984) The innervation of the mammalian pineal gland with special reference to central pinealopetal projections. Pineal Res Rev 2:41–86

Korf HW, Møller M (1985) The central innervation of the mammalian pineal organ. In: Mess B, Ruzsas C, Tima L, Pévet P (eds) Current state of pineal research. Akadémiai Kiadó, Budapest, pp 47–69

Korf HW, Oksche A (1986) The pineal organ. In: Pang PKT, Schreibman M (eds) Vertebrate endocrinology. Morphological considerations. Academic, Orlando, pp 105–145 (Fundamentals and biomedical implications, vol 1)

Korf HW, Vigh-Teichmann I (1984) Sensory and central nervous elements in the avian pineal organ. Ophthalmic Res 16:96–101

Korf HW, Wagner U (1980) Evidence for a nervous connection between the brain and the pineal organ in the guinea pig. Cell Tissue Res 209:505–510

Korf HW, Wagner U (1981) Nervous connections of the parietal eye in adult Lacerta s. sicula Rafinesque as demonstrated by anterograde and retrograde transport of horseradish peroxidase. Cell Tissue Res 219:567–584

Korf HW, Wicht H (1991) Das Pinealorgan der Wirbeltiere: ein Modell für Untersuchungen von Rezeptor- und Effektormechanismen in neuronalen Systemen. Naturwissenschaften 78:437–444

Korf HW, Wicht H (1992) Receptor and effector mechanisms in the pineal organ. Prog Brain Res 91:285–297

Korf HW, Liesner R, Meissl H, Kirk A (1981) Pineal complex of the clawed toad, Xenopus laevis Daud.: structure and function. Cell Tissue Res 216:113–130

Korf HW, Zimmerman NH, Oksche A (1982) Intrinsic neurons and neural connections of the pineal organ of the house sparrow, Passer domesticus, as revealed by anterograde and retrograde transport of horseradish peroxidase. Cell Tissue Res 222:243–260

Korf HW, Foster RG, Ekström P, Schalken JJ (1985a) Opsin-like immunoreaction in the retinae and pineal organs of four mammalian species. Cell Tissue Res 242:645–648

Korf HW, Møller M, Gery I, Zigler JS, Klein DC (1985b) Immunocytochemical demonstration of retinal S-antigen in the pineal organ of four mammalian species. Cell Tissue Res 239:81–85

Korf HW, Oksche A, Ekström P, Veen T van, Zigler JS, Gery I, Stein P, Klein DC (1986a) S-antigen immunocytochemistry. In: O'Brien P, Klein DC (eds) Pineal and retinal relationships. Academic, New York, pp 343–355

Korf HW, Oksche A, Ekström P, Zigler JS, Gery I, Klein DC (1986b) Pinealocyte projections into the mammalian brain revealed with S-antigen antiserum. Science 231:735–737

Korf HW, Sato T, Oksche A (1990) Complex relationships between the pineal organ and the medial habenular nucleus-pretectal region of the mouse as revealed by S-antigen immunocytochemistry. Cell Tissue Res 261:493–500

Korf HW, White BH, Schaad DC, Klein DC (1992) Recoverin in pineal organs and retinae of various vertebrate species including man. Brain Res 595:57–66

Korf HW, Schomerus C, Maronde E, Stehle JH (1996) Signal transduction molecules in the rat pineal organ: Ca^{2+}, pCREB, and ICER. Naturwissenschaften 83:535–543

Kornhauser JM, Nelson DE, Mayo KE, Takahashi JS (1990) Photic and circadian regulation of c-fos gene expression in the hamster suprachiasmatic nucleus. Neuron 5:127–134

Kramm CM, De Grip WJ, Korf HW (1993) Rod-opsin immunoreaction in the pineal organ of the pigmented mouse does not indicate the presence of a functional photopigment. Cell Tissue Res 274:71–78

Kroeber S, Schomerus C, Korf HW (1997) Calcium oscillations in a subpopulation of S-antigen-immunoreactive pinealocytes of the rainbow trout (Oncorhynchus mykiss). Brain Res 744:68–76

Kus L, Handa RJ, McNulty JA (1993) Characterization of a [^{3}H] glutamate binding site in rat pineal gland: enhanced affinity following superior cervical ganglionectomy. J Pineal Res 14:39–44

Kus L, Handa RJ, McNulty JA (1994) Glutamate inhibition of the adrenergic-stimulated production of melatonin in rat pineal gland in vitro. J Neurochem 62:2241–2245

Laitinen JT, Saavedra JM (1990) Characterization of melatonin receptors in the rat suprachiasmatic nuclei: modulation of affinity with cations and guanine nucleotides. Endocrinology 126:2110–2115

Laitinen JT, Laitinen KSM, Kokkola T (1995) Cholinergic signaling in the rat pineal gland. Cell Mol Neurobiol 15:177–192

Larsen PJ, Møller M, Mikkelsen JD (1991) Efferent projections from the periventricular and medial parvocellular subnuclei of the hypothalamic paraventricular nucleus to the circumventricular organs of the rat. A Phaseolus vulgaris leucoagglutinin (PHAL) tracing study. J Comp Neurol 306:462–479

Le Gros Clark WE (1940) The nervous and vascular relations of the pineal gland. J Anat 74:470–494

Lerner AB, Case JD, Takahashi Y, Lee Y, Mori W (1958) Isolation of melatonin, the pineal gland factor that lightens melanocytes. J Am Chem Soc 80:2587

Lerner AB, Case JD, Heinzelmann RV (1959) Structure of melatonin. J Am Chem Soc 81:6084–6085

Lerner AB, Case JD, Takahashi Y (1960) Isolation of melatonin and 5-methoxyindole-3-acetic acid from bovine pineal gland. J Biol Chem 235:1992–1997

Letz B, Schomerus C, Maronde E, Korf HW, Korbmacher C (1997) Stimulation of a nicotinic ACh receptor causes depolarization and activation of L-type Ca^{2+} channels in rat pinealocytes. J Physiol (Lond) 499:329–340

Lewy AJ, Ahmed S, Jackson JM (1992) Melatonin shifts human circadian rhythms according to a phase-response curve. Chronobiol Int 9:380–392

Ling EA, Tan SH, Wong WC (1990) Synaptic junctions between sympathetic axon terminals and pinealocytes in the monkey, Macaca fascicularis. Anat Embryol (Berl) 182:21–27

Liu C, Weaver DR, Jin X, Shearman LP, Pieschl RL, Gribkoff VK, Reppert SM (1997) Molecular dissection of two distinct actions of melatonin on the suprachiasmatic circadian clock. Neuron 19:91–102

Lohse, MJ, Benovic JL, Codina J, Caron MG, Lefkowitz RJ (1990) β-Arrestin: a protein that regulates β-adrenergic receptor function. Science 248:1547–1550

Lovenberg W, Jequier E, Sjoerdsma A (1967) Tryptophan hydroxylation: measurement in pineal gland, brain stem, and carcinoid tumor. Science 155:217–219

Lundberg JM, Terenius L, Hökfelt T, Martling CR, Tatemoto K, Mutt V, Polak J, Bloom S, Goldstein M (1982) Neuropeptide Y-like immunoreactivity in peripheral noradrenergic neurons and effects of NPY on sympathetic function. Acta Physiol Scand 116:477–480

Maestroni GJM, Conti A (1991) Anti-stress role of the melatonin-immuno-opioid network – evidence for a physiological mechanism involving T-cell-derived, immunoreactive beta-endorphin and met-enkephalin binding to thymic opioid receptors. Int J Neurosci 61:289–298

Maronde E, Middendorff R, Telgmann R, Müller D, Hemmings B, Tasken K, Olcese J (1997a) Melatonin synthesis in the bovine pineal gland is regulated by type II cyclic AMP-dependent protein kinase. J Neurochem 68:770–777

Maronde E, Wicht H, Tasken K, Olcese J, Korf HW (1997b) Melatonin biosynthesis in the rat pineal gland is regulated via cAMP-dependent protein kinase type II and phosphorylation of the transcription factor CREB (submitted)

Martin JE, Klein DC (1976) Melatonin inhibition of the neonatal pituitary response to luteinizing hormone-releasing factor. Science 191:301–302

Martin C, Meissl H (1992) Effects of dopaminergic and noradrenergic mechanisms on the neuronal activity of the isolated pineal organ of the trout, Oncorhynchus mykiss. J Neural Transm 88:37–51

Masquilier D, Sassone-Corsi P (1992) Transcriptional cross-talk: nuclear factors CREM and CREB bind to AP-1 sites and inhibit activation by JUN. J Biol Chem 267:22460–22466

Masuo Y, Ohtaki T, Masuda Y, Tsuda M, Fujino M (1992) Binding sites for pituitary adenylate cyclase activating polypeptide (PACAP): comparison with vasoactive intestinal polypeptide (VIP) binding sites location in rat brain sections. Brain Res 575:113–123

Mato E, Santisteban P, Viader M, Capellá G, Fornas O, Puig-Domingo M, Webb SM (1993) Expression of somatostatin in rat pineal cells in culture. J Pineal Res 15:43–45

Matsuura T, Sano Y (1983) Distribution of monoamine-containing nerve fibers in the pineal organ of untreated and sympathectomized dogs. Cell Tissue Res 234:519–531

Matsuura T, Kawata M, Yamada H, Kojima M, Sano Y (1983) Immunohistochemical studies on the peptidergic nerve fibers in the pineal organ of the dog. Arch Histol Jpn 46:373–379

Max M, Menaker M (1992) Regulation of melatonin production by light, darkness, and temperature in the trout pineal. J Comp Physiol [A] 170:479–489

Max M, McKinnon PJ, Seidenman KJ, Barrett RK, Applebury ML, Takahashi JS, Margolskee RF (1995) Pineal opsin: a nonvisual opsin expressed in chick pineal. Science 267:1502–1506

McArthur AJ, Gilette MU, Prosser RA (1991) Melatonin directly resets the rat suprachiasmatic circadian clock in vitro. Brain Res 565:158–161

McCord, CP, Allen FB (1917) Evidence associating pineal gland function with alterations in pigmentation. J Exp Zool 23:207–224

McNulty (1984) Functional morphology of the pineal complex in cyclostomes, elasmobranchs, and bony fishes. Pineal Res Rev 2:1–40

McNulty J, Rathbun WE, Druse MJ (1988) Ultrastructural and biochemical responses of photoreceptor pinealocytes to light and dark in vivo and in vitro. Life Sci 43:845–850

McNulty JA, Kus L, Ottersen OP (1992) Immunocytochemical and circadian biochemical analysis of neuroactive amino acids in the pineal gland of the rat: effect of superior cervical ganglionectomy. Cell Tissue Res 269:515–523

Meiniel (1981) New aspects of the phylogenetic evolution of sensory cell lines in the vertebrate pineal complex. In: Oksche A, Pévet P (eds) The pineal organ: photobiology-biochronometry-endocrinology. Elsevier, Amsterdam, pp 27–48

Meissl H, Dodt E (1981) Comparative physiology of pineal photoreceptor organs. Dev Endocrinol 14:61–80

Meissl H, Donley CS (1980) Change of threshold after light-adaptation of the chromatic response of the frog's pineal organ (stirnorgan). Vision Res 20:379–383

Meissl H, Ekström P (1993) Extraretinal photoreception by pineal systems: a tool for photoperiodic time measurement? Trends Comp Biochem Physiol 1:1223–1240

Meissl H, George SR (1984a) Electrophysiological studies on neuronal transmission in the frog's photosensory pineal organ. The effect of amino acids and biogenic amines. Vision Res 24:1727–1734

Meissl H, George SR (1984b) Photosensory properties of the pineal organ. Microiontophoretic application of excitatory amino acids onto pineal neurons. Ophthalmic Res 16:114–118

Meissl H, Ueck M (1980) Extraocular photoreception of the pineal gland of the aquatic turtle, Pseudemys scripta elegans. J Comp Physiol 140:173–179

Meissl H, Yáñez J (1996) Diazepam increases melatonin secretion of photosensitive pineal organs of trout in the photopic and mesopic range of illumination. Neurosci Lett 207:37–40

Meissl H, Donley CS, Wissler JH (1978) Free amino acids and amines in the pineal organ of the rainbow trout (Salmo gairdneri): influence of light and dark. Comp Biochem Physiol 61 C:401–405

Meissl H, Kroeber S, Yáñez J, Korf HW (1996) Regulation of melatonin production and intracellular calcium concentrations in the trout pineal organ. Cell Tissue Res 286:315–323

Menaker M, Oksche A (1974) The avian pineal organ. In: Farner DS, King JR (eds) Avian biology. Academic, New York, pp 79–118

Meunier AC, Voisin P, Van Camp G, Cenatiempo Y, Müller JM (1991) Molecular characterization and peptide specificity of two vasoactive intestinal peptide (VIP) binding sites in the chicken pineal. Neuropeptides 19:1–18

Mikkelsen JD, Møller M (1988) Vasoactive intestinal peptide in the hypothalamohypophysial system of the mongolian gerbil. J Comp Neurol 273:87–98

Mikkelsen JD, Møller M (1990) A direct neural projection from the intergeniculate leaflet of the lateral geniculate nucleus to the deep pineal gland of the rat, demonstrated with Phaseolus vulgaris leucoagglutinin. Brain Res 520:342–346

Mikkelsen JD, Korf HW, Møller M (1987) Vasoactive intestinal peptide (VIP) in the pineal gland of the rat. In: Trentini GP, de Gaetani C, Pévet P (eds) Fundamentals and clinics in pineal research. Raven, New York, pp 87–90

Mikkelsen JD, Cozzi B, Møller M (1991) Efferent projections from the lateral geniculate nucleus to the pineal complex of the mongolian gerbil (Meriones unguiculatus). Cell Tissue Res 264:95–102

Miyata A, Arimura A, Dahl RR, Minamino N, Uehara A, Jiang L, Culler MD, Coy DH (1989) Isolation of a novel 38 residue hypothalamic polypeptide which stimulates adenylate cyclase in pituitary cells. Biochem Biophys Res Commun 164:567–574

Molina C, Foulkes NS, Lalli E, Sassone-Corsi P (1993) Inducibility and negative autoregulation of CREM: an alternative promoter directs the expression of ICER, an early response repressor. Cell 75:1–20

Møller M (1978) Presence of a pineal nerve (nervus pinealis) in the human fetus: a light and electron microscopical study of the innervation of the pineal gland. Brain Res 154:1–12

Møller M (1992) Fine structure of the pinealopetal innervation of the mammalian pineal gland. Microsc Res Tech 21:188–204

Møller M, Nielsen JT, van Veen T (1979) Effect of superior cervical ganglionectomy on monoamines in the epithalamic area of the Mongolian gerbil (Meriones unguiculatus). A fluorescence histochemical study. Cell Tissue Res 201:1–9

Møller M, Mikkelsen JD, Fahrenkrug J, Korf HW (1985) The presence of vasoactive intestinal polypeptide (VIP)-like-immunoreactive nerve fibres and VIP-receptors in the pineal gland of the mongolian gerbil (Meriones unguiculatus). An immunohistochemical and receptor-autoradiographic study. Cell Tissue Res 241:333–340

Møller M, Mikkelsen JD, Martinet L (1990) Innervation of the mink pineal with neuropeptide Y (NPY)-containing nerve fibers. An experimental immunohistochemical study. Cell Tissue Res 261:477–483

Møller M, Mikkelsen JD, Holst JJ, Phansuwan-Pujito P (1992) Somatostatin and prosomatostatin immunoreactive nerve fibers in the bovine pineal gland. Neuroendocinology 56:278–283

Møller M, Phansuwan-Pujito P, Govitrapong P, Schmidt P (1993) Indications for a central innervation of the bovine pineal gland with substance P-immunoreactive nerve fibers. Brain Res 611:347–351

Møller M, Phansuwan-Pujito P, Pramanlkijja S, Kotchabhakdi N, Govitrapong P (1994) Innervation of the cat pineal gland by neuropeptide Y-immunoreactive nerve fibers: an experimental immunohistochemical study. Cell Tissue Res 276:545–550

Møller M, Phansuwan-Pujito P, Morgan KC, Badiu C (1997) Localization and diurnal expression of mRNA encoding the β1-adrenergic receptor in the rat pineal gland: an in situ hybridization study. Cell Tissue Res 288:279–284

Møllgard K, Møller M (1973) On the innervation of the human fetal pineal gland. Brain Res 52:428–432

Montminy MR, Bilezikjian LM (1987) Binding of a nuclear protein to the cyclic AMP response element of the somatostatin gene. Nature 328:175–178

Moore RY, Sibony P (1988) Enkephalin-like immunoreactivity in neurons in the human pineal gland. Brain Res 457:395–398

Morgan JI, Curran T (1991) Stimulus-transcription coupling in the nervous system: involvement of the inducible proto-oncogenes fos and jun. Annu Rev Neurosci 14:421–451

Morgan PJ, Williams LM, Lawson W, Riddoch G (1988) Adrenergic and VIP stimulation of cyclic AMP accumulation in ovine pineals. Brain Res 447:279–286

Morgan PJ, Davidson GR, Lawson W (1989) Evidence for dual adrenergic receptor regulation of ovine pineal function. J Pineal Res 7:175–183

Morgan PJ, King TP, Lawson W, Slater D, Davidson G (1991a) Ultrastructure of melatonin-responsive cells in the ovine pars tuberalis. Cell Tissue Res 263:529–534

Morgan PJ, Lawson W, Davidson G (1991b) Interaction of forskolin and melatonin on cyclic AMP generation in pars tuberalis cells of ovine pituitary. J Neuroendocr 3:497–501

Morgan PJ, Barrett P, Howell HE, Helliwell R (1994) Melatonin receptors: localization, molecular pharmacology and physiological significance. Neurochem Int 24:101–146

Morita Y (1966) Entladungsmuster pinealer Neurone der Regenbogenforelle (Salmo irideus) bei Belichtung des Zwischenhirns. Pflugers Arch 289:155–167

Morita Y, Dodt E (1965) Nervous activity of the frog's epiphysis cerebri in relation to illumination. Experientia 21:221–222

Morita Y, Tabata M, uchida K, Samejima M (1992) Pineal-dependent locomotor activity of lamprey, Lampetra japonica, measured in relation to LD cycle and circadian rhythmicity. J Comp Physiol 171:555–562

Moriyama Y, Yamamoto A (1995a) Microvesicles isolated from bovine pineal gland specifically accumulate L-glutamate. FEBS Lett 367:233–236

Moriyama Y, Yamamoto A (1995b) Vesicular L-glutamate transporter in microvesicles from bovine pineal glands. J Biol Chem 270:22314–22320

Moujir F, Reiter RJ, Rodriguez C, Yaga K (1992) β-adrenergic and peptide N-terminal histidine and C-terminal isoleucine stimulation of N-acetyltransferase activity and melatonin production in the cultured rat pineal gland. Endocrinology 130:2076–2082

Nielsen JT, Møller M (1978) Innervation of the pineal gland in the Mongolian gerbil (Meriones unguiculatus). A fluorescence microscopical study. Cell Tissue Res 187:235–250

Nürnberger F, Korf HW (1981) Oxytocin- and vasopressin-immunoreactive nerve fibers in the pineal gland of the hedgehog, Erinaceus europaeus L. Cell Tissue Res 220:87–97

Okano T, Yoshizawa T, Fukada Y (1994) Pinopsin is a chicken pineal photoreceptive molecule. Nature 372:94–96

Oksche A (1965) Survey of the development and comparative morphology of the pineal organ. Prog Brain Res 10:3–29

Oksche A (1971) Sensory and glandular elements of the pineal organ. In: Wolstenholme GEW, Knight J (eds) The pineal gland. Churchill-Livingstone, Edinburgh, pp 127–146

Oksche A, Hartwig HG (1979) Pineal sense organs – components of photoneuroendocrine systems. Prog Brain Res 52:113–130

Oksche A, Kirschstein H (1967) Die Ultrastruktur der Sinneszellen im Pinealorgan von Phoxinus laevis. Z Zellforsch 78:151–166

Oksche A, Kirschstein H (1968) Unterschiedlicher elektronenmikroskopischer Feinbau der Sinneszellen im Parietalauge und im Pinealorgan (Epiphysis cerebri) der Lacertilia. Ein Beitrag zum Epiphysenproblem. Z Zellforsch 87:159–192

Oksche A, Kirschstein H (1971) Weitere elektronenmikroskopische Untersuchungen am Pinealorgan von Phoxinus laevis (Teleostei, Cyprinidae). Z Zellforsch 112:572–588

Oksche A, Korf HW, Rodríguez EM (1987) Pinealocytes as photoneuroendocrine units of neuronal origin: concepts and evidence. Adv Pineal Res 2:1–18

Olcese J (1991) Neuropeptide Y: an endogenous inhibitor of norepinephrine-stimulated melatonin secretion in the rat pineal gland. J Neurochem 57:943–947

Ostrowski NL, Lolait SJ, Young III, WS (1994) Cellular localization of vasopressin V1a receptor messenger ribonucleic acid in adult male rat brain, pineal, and brain vasculature. Endocrinology 135:1511–1528

Owman C (1964) Sympathetic nerves probably storing two types of monoamines in the rat pineal gland. Int J Neuropharmacol 3:105–112

Owman C (1965) Localization of neuronal and parenchymal monoamines under normal and experimental conditions in the mammalian pineal gland. Prog Brain Res 10:423–453

Owman C, Rüdeberg C (1970) Light, fluorescence, and electron microscopic studies on the pineal organ of the pike, Esox lucius L., with special regard to 5-hydroxytryptamine. Z Zellforsch 107:522–550

Owman C, Rüdeberg C, Ueck M (1970) Fluoreszenzmikroskopischer Nachweis biogener Monoamine in der Epiphysis cerebri von Rana esculenta und Rana pipiens. Z Zellforsch 111:550–558

Pangerl B, Pangerl A, Reiter RJ (1990) Circadian variations of adrenergic receptors in the mammalian pineal gland: a review. J Neural Transm 81:17–29

Paul E, Hartwig HG, Oksche A (1971) Neurone und zentralnervöse Verbindungen des Pinealorgans der Anuren. Z Zellforsch 112:466–493

Pellegrino De Iraldi A, Zieher LM, De Robertis E (1965) Ultrastructure and pharmacological studies of nerve endings in the pineal organ. Prog Brain Res 10:389–422

Pévet P (1985) 5-Methoxyindoles, pineal, and seasonal reproduction – a new approach. In: Mess B, Ruzsas C, Tima L, Pévet P (eds) The pineal gland: Current state of pineal research. Akadémiai Kiadó, Budapest, pp 163–186

Pfeffer M, Stehle JH, Schloss P, Betz H, Korf HW (1997) Nachweis eines Serotonintransporters in Ganglion cervicale superius und sympathischer Innervation des Pinealorgans der Ratte durch in situ Hybridisierung und Immunzytochemie. Ann Anat 179 [Suppl]:338

Phansuwan-Pujito P, Mikkelsen JD, Govitrapong P Møller M (1991) A cholinergic innervation of the bovine pineal gland visualized by immunohistochemical detection of choline acetyltransferase-immunoreactive nerve fibers. Brain Res 545:49–58

Pines G (1927) Über die Innervation der Epiphyse. Z Ges Neurol 11:365–369

Pratt BL, Takahashi JS (1987) Alpha-2 adrenergic regulation of melatonin release in chick pineal cell cultures. J Neurosci 7:3665–3674

Pratt JM, Takahashi JS (1989) Vasoactive intestinal polypeptide and α_2-adrenoreceptor agonists regulate adenosine 3',5'-monophosphate accumulation and melatonin release in chick pineal cell cultures. Endocrinology 125:2375–2384

Quay WB (1963) Circadian rhythm in rat pineal serotonin and its modifications by estrous cycle and photoperiod. Gen Comp Endocrinol 3:473–479

Quay WB (1974) Pineal chemistry. Thomas, Springfield

Quay WB, Kappers JA, Jongkind JF (1968) Innervation and fluorescence histochemistry of monoamines in the pineal organ of a snake (Natrix natrix). J Neuro Visc Relat 31:11–25

Redecker P (1993) Synaptophysin: ein Membranprotein synaptischer Vesikel im Nervensystem und synapsenähnlicher Mikrovesikel in neuroendokrinen Zellen. Thesis, Faculty of Medicine, Hannover, Germany

Redecker P (1995) The ras-like rab3 A protein is present in pinealocytes of the gerbil pineal gland. Neurosci Lett 184:117–120

Redecker P (1996) Synaptotagmin I, synaptobrevin II, and syntaxin I are coexpressed in rat and gerbil pinealocytes. Cell Tissue Res 283:443–454

Redecker P, Veh RW (1994) Glutamate immunoreactivity is enriched over pinealocytes of the gerbil pineal gland. Cell Tissue Res 278:579–588

Redecker P, Grube D, Jahn R (1990) Immunohistochemical localization of synaptophysin (p38) in the pineal gland of the Mongolian gerbil (Meriones unguiculatus). Anat Embryol (Berl) 181:433–440

Reiter RJ (1991) Pineal melatonin: cell biology of its synthesis and of its physiological interactions. Endocrin Rev 12:151–180

Reppert SM, Weaver DR (1997) Forward genetic approach strikes gold: cloning of a mammalian clock gene. Cell 89:487–490

Reppert SM, Weaver DR, Rivkees SA, Stopa EG (1988) Putative melatonin receptors in a human biological clock. Science 242:78–91

Reppert SM, Weaver DR, Ebisawa T (1994) Cloning and characterization of a mammalian melatonin receptor that mediates reproductive and circadian responses. Neuron 13:1177–1185

Reppert SM Godson C, Mahle CD, Weaver DR, Slaugenhaupt SA, Gusella JF (1995a) Molecular characterization of a second melatonin receptor expressed in the human retina and brain: the Mel_{1b} melatonin receptor. Proc Natl Acad Sci USA 92:8734–8738

Reppert SM, Weaver DR, Cassone VM, Godson C, Kolakowski LF (1995b) Melatonin receptors are for the birds: molecular analysis of two receptor subtypes differentially expressed in chick brain. Neuron 15:1003–1015

Reuss S, Moore RY (1989) Neuropeptide Y-containing neurons in the rat superior cervical ganglion: projections to the pineal gland. J Pineal Res 6:307–316

Reuss S, Schröder H (1987) Neuropeptide Y effects on pineal melatonin synthesis in the rat. Neurosci Lett 74:158–162

Reuss S, Schröder H (1988) Principal neurons projecting to the pineal gland in close association with small intensely fluorescent cells in the superior cervical ganglion of rats. Cell Tissue Res 254:97–100

Reuss S, Schröder B, Schröder H, Maelicke A (1992) Nicotinic cholinoceptors in the rat pineal gland as analysed by western blot, light- and electron microscopy. Brain Res 573:114–118

Robertson LM, Takahashi JS (1988) Circadian clock in cell culture: I. Oscillation of melatonin release from dissociated chick pineal cells in flow-through microcarrier culture. J Neurosci 8:12–21

Roca AL, Godson C, Weaver DR, Reppert SM (1996) Structure, characterization, and expression of the gene encoding the mouse Mel_{1a} melatonin receptor. Endocrinology 137:3469–3477

Rodriguez IR, Mazuruk K, Schoen TJ, Chader GJ (1995) Structural analysis of the human hydroxyindole-O-methyltransferase gene. J Biol Chem 269:31969–31977

Rollag MD (1988) Response of amphibian melanophores to melatonin. Pineal Res Rev 6:67–93

Romero JA, Zatz M, Kebabian JW, Axelrod J (1975) Circadian cycles in binding of ^{3}H-alprenolol to β-adrenergic receptor sites in rat pineal. Nature 258:435–436

Romijn HJ (1973) Structure and innervation of the pineal gland of the rabbit, Oryctolagus cuniculus (L.). I. A light microscopic investigation. Z Zellforsch 139:473–485

Romijn HJ (1975) Structure and innervation of the pineal gland of the rabbit, Oryctolagus cuniculus (L.). III. An electron microscopic investigation of the innervation. Cell Tissue Res 157:25–51

Ronnekleiv OK (1988) Distribution in the macaque pineal of nerve fibers containing immunoreactive substance P, vasopressin, oxytocin, and neurophysins. J Pineal Res 5:259–271

Ronnekleiv OK, Kelly MJ (1984) Distribution of substance P neurons in the epithalamus of the rat: an immunohistochemical investigation. J Pineal Res 1:355–370

Roseboom PH, Klein DC (1995) Norepinephrine stimulation of pineal cyclic AMP response element-binding protein phosphorylation: involvement of a β-adrenergic/cyclic AMP mechanism. Mol Pharmacol 47:439–449

Roseboom PH, Coon SL, Baler R, McCune SK, Weller JL, Klein DC (1996) Melatonin synthesis: analysis of the more than 150-fold nocturnal increase in serotonin N-acetyltransferase messenger ribonucleic acid in the rat pineal gland. Endocrinology 137:3033–3044

Saavedra JM, Brownstein M, Axelrod J (1973) A specific and sensitive enzymatic-isotopic microassay for serotonin in tissues. J Pharmacol Exp Ther 186:508–515

Sáez JC, Berthoud VM, Kadle R, Traub O, Nicholson BJ, Bennett MVL, Dermietzel R (1991) Pinealocytes in rats: connexin identification and increase in coupling caused by norepinephrine. Brain
Res 568:265–275

Samejima M, Tamotsu S, Matanabe K, Morita Y (1989) Photoreceptor cells and neural elements with
long axonal processes in the pineal organ of the lamprey, Lampetra japonica, identified by use of
the horseradish peroxidase method. Cell Tissue Res 258:219–224

Sarda N, Gharib A, Reynaud D, Ou L, Pacheco H (1989) Identification of adenosine receptor in rat
pineal gland: evidence for A-2 selectivity. J Neurochem 53:733–737

Sasek CA, Zigmond RE (1989) Localization of vasoactive intestinal peptide- and peptide histidine
isoleucine amide-like immunoreactivities in the rat superior cervical ganglion and its nerve trunks.
J Comp Neurol 280:522–523

Sato T, Wake K (1983) Innervation of the pineal organ. Cell Tissue Res 233:237–264

Sato T, Wake K (1984) Regressive post-hatching development of acetylcholinesterase-positive neurons in the pineal organ of Coturnix coturnix japonica and Gallus gallus. Cell Tissue Res
237:267–275

Sato T, Wake K, Kramm C, Korf HW (1990) Chicken pineal organs during post-hatching development:
photoreceptor-specific characteristics and innervation. In: Gupta D, Ranke B, Wollmann R (eds)
Neuroendoendocrinology: new frontiers. Brain Research Promotion, Tübingen, pp 191–200

Schaad NC, Vanecek J, Schulz PE (1994) Photoneural regulation of rat pineal nitric synthase. J
Neurochem 62:2496–2499

Schaad NC, Vanecek J, Rodriguez IR, Klein DC, Holtzclaw L, Russell JT (1995) Vasoactive intestinal
peptide elevates pinealocyte intracellular calcium concentrations by enhancing influx: evidence
for involvement of a cyclic GMP-dependent mechanism. Mol Pharmacol 47:923–933

Scharrer E (1928) Die Lichtempfindlichkeit blinder Elritzen. Untersuchungen über das Zwischenhirn
der Fische. Z Vergl Physiol 7:1–38

Scharrer E (1964) Photo-neuro-endocrine systems: General concepts. Ann NY Acad Sci 117:13–22

Schomerus C, Ruth P, Korf HW (1994) Photoreceptor-specific proteins in the mammalian pineal
organ: immunocytochemical data and functional considerations. Acta Neurobiol Exp 54
[Suppl]:9–17

Schomerus C, Laedtke E, Korf HW (1995) Calcium responses of isolated, immunocytochemically
identified rat pinealocytes to noradrenergic, cholinergic and vasopressinergic stimulations.
Neurochem Int 27:163–175

Schomerus C, Maronde E, Laedtke E, Korf HW (1996) Vasoactive intestinal peptide (VIP) and
pituitary adenylate cyclase-activating polypeptide (PACAP) induce phosphorylation of the transcriptional factor CREB in subpopulations of rat pinealocytes: immunocytochemical and immunochemical evidence. Cell Tissue Res 286:305–313

Schon F, Allen JM, Yeats JC, Allen YS, Ballesta J, Polak JM, Kelly JS, Bloom SR (1985) Neuropeptide
Y innervation of the rodent pineal gland and cerebral vessels. Neurosci Lett 57:65–71

Schröder H (1986) Neuropeptide Y (NPY)-like immunoreactivity in peripheral and central nerve
fibres of the golden hamster (Mesocricetus auratus) with special respect to pineal gland innervation. Histochemistry 85:321–325

Schröder H, Vollrath L (1985) Distribution of dopamine-beta-hydroxylase-like immunoreactivity in
the rat pineal organ. Histochemistry 83:375–380

Schröder H, Vollrath L (1986) Neuropeptide Y (NPY)-like immunoreactivity in the guinea pig pineal
organ. Neurosci Lett 63:285–289

Schröder H, Reuss S, Stehle J, Vollrath L (1988) Intraarterially administered vasopressin inhibits
nocturnal pineal melatonin synthesis in the rat. Comp Biochem Physiol 89 A:651–653

Sheng M, Greenberg ME (1990) The regulation and function of c-fos and other immediate early genes
in the nervous system. Neuroscience 4:477–485

Sheridan MN, Sladek JR (1975) Histofluorescence and ultrastructural analysis of hamster and monkey
pineal. Cell Tissue Res 164:145–152

Shiotani Y, Yamano M, Shiosaka S, Emson PC, Hillyard CJ, Girgis S, MacIntyre I (1986) Distribution
and origins of substance P (SP)-, calcitonin gene-related peptide (CGRP)-, vasoactive intestinal
polypeptide (VIP)- and neuropeptide Y (NPY)-containing nerve fibres in the pineal gland of
gerbils. Neurosci Lett 70:187–192

Simonneaux V, Ouichou A, Burbach JPH, Pévet P (1990) Vasopressin and oxytocin modulation of melatonin secretion from rat pineal glands. Peptides 11:1075–1079

Simonneaux V, Ouichou A, Pévet P (1993) Pituitary adenylate cyclase-activating polypeptide (PA-CAP) stimulates melatonin synthesis from the rat pineal gland. Brain Res 603:148–152

Siuciak JA, Fang JM, Dubocovich ML (1990) Autoradiographic localization of 2-[^{125}I]-iodomelatonin binding sites in the brains of C3H/HeN and C57BL/6 J strains of mice. Eur J Pharmacol 180:387–390

Snyder SH, Axelrod J, Wurtman RJ, Fischer E (1965) Control of 5-hydroxytryptophan decarboxylase activity in the rat pineal gland by sympathetic nerves. J Pharmacol Exp Ther 147:371–375

Solessio E, Engbretson GA (1993) Antagonistic chromatic mechanisms in photoreceptors of the parietal eye of lizards. Nature 364:442–445

Spessert R, Layes E, Vollrath L (1993) Adrenergic stimulation of cyclic GMP formation requires NO-dependent activation of cytosolic guanylate cyclase in rat pinealocytes. J Neurochem 61:138–143

Stankov B, Cimino M, Marini P, Lucini V, Fraschini F, Clementi F (1993) Identification and functional significance of nicotinic cholinergic receptors in the rat pineal gland. Neurosci Lett 156:131–134

Stehle J (1990) Melatonin binding sites in brain of 2-day old chicken: an autoradiographical localization. J Neural Transm 81:83–89

Stehle JH (1995) Pineal gene expression: dawn in a dark matter. J Pineal Res 18:179–190

Stehle J, Vanecek J, Vollrath L (1989) Effects of melatonin on spontaneous electrical activity of neurons in the rat suprachiasmatic nuclei: an iontophoretic study. J Neural Transm 78:173–177

Stehle J Reuss S, Riemann R, Seidel A, Vollrath L (1991) The role of arginine-vasopressin for pineal melatonin synthesis in the rat: involvement of vasopressinergic receptors. Neurosci Lett 123:131–134

Stehle JH, Foulkes NS, Molina CA, Simonneaux V, Pévet P, Sassone-Corsi P (1993) Adrenergic signals direct rhythmic expression of transcriptional repressor CREM in the pineal gland. Nature 356:314–320

Stehle JH, Foulkes NS, Pévet P, Sassone-Corsi P (1995) Developmental maturation of pineal gland function: Synchronized CREM inducibility and adrenergic stimulation. Mol Endocrinol 9:706–716

Stehle JH, Pfeffer M, Kühn R, Korf HW (1996) Light-induced expression of transcription factor ICER (inducible cAMP early repressor) in rat suprachiasmatic nucleus is phase-restricted. Neurosci Lett 217:169–172

Stehle JH, Pfeffer M, Krug L, Korf HW (1997) β-adrenergic receptors in rat pineal gland: diurnal and ontogenetic expression and regulation. Ann Anat 179 [Suppl]:15

Studnicka FK (1905) Die Parietalorgane. In: Oppel A (ed) Lehrbuch der vergleichenden mikroskopischen Anatomie, vol 5. Fischer, Jena, pp 1–254

Subhedar N, Cerda J, Wallace RA (1996) Neuropeptide Y in the forebrain and retina of the killifish, Fundulus heroclitus. Cell Tissue Res 283:313–323

Sugden D (1989) Melatonin biosynthesis in the mammalian pineal gland. Experientia 45:922–932

Sugden D (1990) 5-hydroxytryptamine amplifies β-adrenergic stimulation of N-acetyltransferase activity in rat pinealocytes. J Neurochem 55:1655–1658

Sugden D, Klein DC (1983) β-adrenergic receptor control of rat hydroxyindole-O-methyltransferase. Endocrinology 113:348–353

Sugden D, Klein DC (1985) Development of the rat pineal α_1-adrenoceptor. Brain Res 325:345–348

Sugden D, Klein DC (1987) A cholera toxin substrate regulates cyclic GMP content of rat pinealocytes. J Biol Chem 262:7447–7450

Takahashi JS (1994) ICER is nicer at night, sir! Curr Biol 4:165–168

Takahashi JS (1996) Ion channels get the message. Nature 382:117–118

Takahashi JS, Hamm H, Menaker M (1980) Circadian rhythm of melatonin release from individual superfused chicken pineal glands in vitro. Proc Natl Acad Sci USA 77:2319–2322

Takahashi JS, Murakami N, Nikaido S, Pratt B, Robertson LM (1989) The avian pineal – a vertebrate model system of the circadian oscillator: cellular regulation of circadian rhythms by light, secondary messengers, and macromolecular synthesis. Rec Prog Horm Res 45:279–352

Tamotsu S, Korf HW, Morita Y, Oksche A (1990) Immunocytochemical localization of serotonin and photoreceptor-specific proteins (rod-opsin, S-antigen) in the pineal complex of the river lamprey, Lampetra japonica, with special reference to photoneuroendocrine cells. Cell Tissue Res 262:205–216

Tamotsu S, Schomerus C, Stehle JH, Roseboom PH, Korf HW (1995) Norepinephrine-induced phosphorylation of the transcription factor CREB in isolated rat pinealocytes: an immunocyto-chemical study. Cell Tissue Res 282:219–226

Taugner R, Schiller A, Rix E (1981) Gap junction between pinealocytes. A freeze-fracture study of the pineal gland in rats. Cell Tissue Res 218:303–314

Thibault C, Falcón J, Greenhouse SS, Lowery A, Gern WA, Collin JP (1993) Regulation of melatonin production by pineal photoreceptor cells: role of cyclic nucleotides in the trout (Oncorhynchus mykiss). J Neurochem 6:332–339

Trueman T, Herbert J (1970) Monoamines and acetylcholinesterase in the pineal gland and habenula of the ferret. Z Zellforsch 109:83–100

Tsuchiya M, Kaku K, Matsuda M, Kaneko T, Yanaihara N (1987) Demonstration of receptors specific for peptide N-terminal histidine and C-terminal isoleucine (PHI) using rat PHI and rat dispersed pineal cells. Biomed Res 8:45–51

Uddman R, Alumets J, Håkanson R, Loren I, Sundler F (1980) Vasoactive intestinal peptide (VIP) occurs in the nerves of the pineal gland. Experientia 36:1119–1120

Ueck M (1970) Weitere Untersuchungen zur Feinstruktur und Innervation des Pinealorgans von Passer domesticus L. Z Zellforsch 105:276–302

Ueck M (1973) Fluoreszenzmikroskopische und elektronenmikroskopische Untersuchungen am Pinealorgan verschiedener Vogelarten. Z Zellforsch 137:37–62

Ueck M (1979) Innervation of the vertebrate pineal. Prog Brain Res 52:45–87

Ueck M, Kobayashi H (1972) Vergleichende Untersuchungen über acetylcholinesterasehaltige Neurone im Pinealorgan der Vögel. Z Zellforsch 129:140–160

Vacas MI, Samiento MIK, Pereyra EN, Etchegoyen GS, Cardinali DP (1987) In vitro effect of neuropeptide Y on melatonin and norepinephrine release in rat pineal gland. Cell Mol Neurobiol 7:309–315

Vanecek J (1988) Melatonin binding sites. J Neurochem 51:1436–1440

Vanecek J, Klein DC (1992a) Melatonin inhibits gonadotropin releasing hormone-induced elevation of intracellular Ca2+ in neonatal rat pituitary cells. Endocrinology 130:701–707

Vanecek J, Klein DC (1992b) Sodium-dependent effects of melatonin on membrane potential of neonatal rat pituitary cells. Endocrinology 131:939–946

Vanecek J, Vollrath L (1989) Melatonin inhibits cyclic AMP and cyclic GMP accumulation in rat pituitary. Brain Res 505:157–159

Vanecek J, Vollrath L (1990) Melatonin modulates diacylglycerol and arachidonic acid metabolism in the anterior pituitary of immature rats. Neurosci Lett 110:199–203

Vanecek J, Sugden D, Weller J, Klein DC (1985) Atypical synergistic α_1- and β_1-adrenergic regulation of adenosine 3',5'-monophosphate and guanosine 3',5'-monophosphate in rat pinealocytes. Endocrinology 116:2167–2173

Vanecek J, Pavlik A, Illnerova H (1987) Hypothalamic melatonin receptor sites revealed by autoradiography. Brain Res 435:359–363

Van Veen T, Östholm T, Gierschik P, Spiegel A, Somers R, Korf HW, Klein DC (1986) Alpha-transducin immunoreactivity in retinae and sensory pineal organs of adult vertebrates. Proc Natl Acad Sci USA 83:912–916

Van Wyk E, Daya S (1994) Glutamate inhibits the isoprenaline-induced raise in melatonin synthesis by organ cultures of rat pineal glands. Med Sci Res 22:635–636

Vigh B, Vigh-Teichmann I, Röhlich P, Aros B (1982) Immunoreactive opsin in the pineal organ of reptiles and birds. Z Mikrosk Anat Forsch 96:113–129

Vigh B, Vigh-Teichmann I, Debreceni K, Takacs J (1995) Similar fine-structural localization of immunoreactive glutamate in the pineal complex and retina of frogs. Arch Histol Cytol 58:37–44

Vigh B, Debreceni K, Fejer Z, Vigh-Teichmann I (1997) Immunoreactive excitatory amino acids in the parietal eye of lizards, a comparison with the pineal organ and retina. Cell Tissue Res 287:275–283

Vigh-Teichmann I, Vigh B (1990) Opsin: immunocytochemical characterization of different types of photoreceptors in the frog pineal organ. J Pineal Res 8:323–333

Vigh-Teichmann I, Korf HW, Oksche A, Vigh B (1982) Opsin-immunoreactive outer segments and acetylcholinesterase-positive neurons in the pineal complex of Phoxinus phoxinus (Teleostei, Cyrinidae). Cell Tissue Res 262:205–216

Vigh-Teichmann I, Petter H, Vigh B (1991) GABA-immunoreactive and -immunonegative intrinsic secondary neurons in the cat pineal organ. J Pineal Res 10:18–29

Voisin P, Collin JP (1986) Regulation of chicken pineal arylalkylamine-N-acetyltransferase by postsynaptic alpha-2-adrenergic receptors. Life Sci 39:2025–2032

Voisin P, Guerlotte J, Collin JP (1988) An antiserum against chicken hydroxyindole-O-methyltransferase reacts with the enzyme from pineal gland and retina and labels pineal modified photoreceptors. Mol Brain Res 4:53–61

Vollrath L (1981) The pineal organ. In: Oksche A, Vollrath L (eds) Handbuch der Mikroskopischen Anatomie des Menschen, vol VI/7. Springer, Berlin Heidelberg New York, pp 1–665

Vollrath L, Schröder H (1987) Neuronal properties of mammalian pinealocytes? In: Trentini GP, de Gaetani C, Pévet P (eds) Fundamentals and clinics in pineal research. Raven, New York, pp 13–23

Von Frisch K (1911) Beiträge zur Physiologie der Pigmentzellen in der Fischhaut. Pflugers Arch 138:319–387

Wake K (1973) Acetylcholinesterase-containing nerve cells and their distribution in the pineal organ of the goldfish, Carassius auratus. Z Zellforsch Mikrosk Anat 145:287–298

Wake K, Ueck M, Oksche A (1974) Acetylcholinesterase-containing nerve cells in the pineal complex and subcommissural area of the frogs, Rana ridibunda and Rana esculenta. Cell Tissue Res 154:423–442

Wartenberg H, Baumgarten HG (1969) Untersuchungen zur fluoreszenz- und elektronenmikroskopischen Darstellung von 5-Hydroxytryptamin (5-HT) im Pinealorgan von Lacerta viridis und L. muralis. Z Anat Entwicklungsgesch 128:185–210

Weaver DR, Rivkees SA, Reppert SM (1989) Localization and characterization of melatonin receptors in rodent brain by in vitro autoradiography. J Neurosci 9:2581–2590

Weaver DR, Provencio I, Carlson LL, Reppert SM (1991) Melatonin receptors and signal transduction in photorefractory Siberian hamsters (Phodopus sungorus). Endocrinology 128:1086–1092

Weaver DR, Stehle JH, Stopa EG, Reppert SM (1993) Melatonin receptors in human hypothalamus and pituitary – implications for circadian and reproductive responses to melatonin. J Clin Endocrinol Metab 76:295–301

Weigle C, Wicht H, Korf HW (1996) A possible homologue of the suprachiasmatic nucleus in the hypothalamus of lampreys (Lampetra fluviatilis L.). Neurosci Lett 217:173–176

Weihe E, Tao-Cheng JH, Schäfer MKH, Erickson JD, Eiden LE (1996) Visualization of the vesicular acetylcholine transporter in cholinergic nerve terminals and its targeting to a specific population of small synaptic vesicles. Proc Natl Acad Sci USA 93:3547–3552

Weissbach H, Redfield BG, Axelrod J (1960) Biosynthesis of melatonin: enzymatic conversion of serotonin to N-acetylserotonin. Biochim Biophys Acta 43:352–353

Weissbach H, Redfield BG, Axelrod J (1961) The enzymatic acetylation of serotonin and other naturally occurring amines. Biochem Biophys Acta 54:190–192

Welsh M (1983) CSF-contacting pinealocytes in the pineal recess of the Mongolian gerbil: a correlative scanning and transmission electron microscopic study. Am J Anat 166:483–493

Welsh DK, Logothetis DE, Meister M, Reppert SM (1995) Individual neurons dissociated from rat suprachiasmatic nucleus express independently phased circadian firing rhythms. Neuron 14:697–706

White B, Sekura RD, Rollag MD (1987) Pertussis toxin blocks melatonin-induced pigment aggregation in Xenopus dermal melanophores. J Comp Physiol 157:153–159

Williams LM, Morgan PJ, Pelletier G, Riddoch GI, Lawson W, Davidson GR (1989) Neuropeptide Y (NPY) innervation of the ovine pineal gland. J Pineal Res 7:345–353

Winters KE, Morrissey JJ, Loos PJ, Lovenberg W (1977) Pineal protein phosphorylation during serotonin N-acetyltransferase induction. Proc Natl Acad Sci USA 74:1928–1931

Wisden W, Seeburg PH (1993) A complex mosaic of high-affinity kainate receptors in rat brain. J Neurosci 13:3582–3598

Wood JG (1973) The effects of niamid and reserpine on the nerve endings of the pineal gland. Z Zellforsch 145:151–166

Wurtman RJ, Anton-Tay F (1969) The mammalian pineal as a neuroendocrine transducer. Rec Prog Horm Res 25:493–522

Yamada H, Yamamoto A, Takahashi M, Michibata H, Kumon H, Moriyama Y (1996a) The L-type Ca^{2+} channel is involved in microvesicle-mediated glutamate exocytosis from rat pinealocytes. J Pineal Res 21:165–174

Yamada H, Yamamoto A, Yodozawa S, Kozaki S, Takahashi M, Morita M, Michibata H, Furuichi T, Mikoshiba K, Moriyama Y (1996b) Microvesicle-mediated exocytosis of glutamate is a novel paracrine-like chemical transduction mechanism and inhibits melatonin secretion in rat pinealocytes. J Pineal Res 21:175–191

Yuwiler A (1983) Vasoactive intestinal peptide stimulation of pineal serotonin-N-acetyltransferase activity: General characteristics. J Neurochem 41:146–153

Yuwiler A (1987) Synergistic action of post-synaptic alpha-receptor stimulation on VIP-induced increases in pineal N-acetyltransferase activtiy. J Neurochem 49:806–811

Yuwiler A, Klein DC, Buda M, Weller JL (1977) Adrenergic control of pineal N-acetyltransferase activity: developmental aspects. Am J Physiol 233:E141–E146

Zachmann A, Knijff SCM, Ali MA, Anctil M (1992) Effects of photoperiod and different intensities of light exposure on melatonin levels in the blood, pineal organ, and retina of the brook trout (Salvelinus fontinalis Mitchill). Can J Zool 70:25–29

Zatz M (1992) Agents that affect calcium influx can change cyclic nucleotide levels in cultured chick pineal cells. Brain Res 583:304–307

Zatz M, Mullen DA (1988a) Norepinephrine, acting via adenylate cyclase, inhibits melatonin output but does not phase-shift the pacemaker in cultured chick pineal cell. Brain Res 450:137–143

Zatz M, Mullen D (1988b) Photoendocrine transduction in cultured chick pineal cells II. Effects of forskolin, 8-bromocyclic AMP, and 8-bromocyclic GMP on the melatonin rhythm. Brain Res 453:51–62

Zatz M, Mullen DA, Moskal JR (1988) Photoendocrine transduction in cultured chick pineal cells: effects of light, dark, and potassium on the melatonin rhythm. Brain Res 438:199–215

Zatz M, Kasper G, Marquez CR (1990) Vasoactive intestinal peptide stimulates chick pineal melatonin production and interacts with other stimulatory and inhibitory agents but does not show α_1-adrenergic potentiation. J Neurochem 55:1149–1153

Zhang ET, Mikkelsen JD, Møller M (1991) Tyrosine hydroxylase- and neuropeptide Y-immuno-reactive nerve fibers in the pineal complex of untreated rats and rats following removal of the superior cervical ganglia. Cell Tissue Res 265:63–71

Zimmerman NH, Menaker M (1979) The pineal: a pacemaker within the circadian system of the house sparrow. Proc Natl Acad Sci USA 76:999–1003

Zweig M, Axelrod J (1969) Relationship between catecholamines and serotonin in sympathetic nerves of the rat pineal gland. J Neurobiol 1:87–97

Subject Index

Springer
and the
environment

At Springer we firmly believe that an international science publisher has a special obligation to the environment, and our corporate policies consistently reflect this conviction.

We also expect our business partners – paper mills, printers, packaging manufacturers, etc. – to commit themselves to using materials and production processes that do not harm the environment. The paper in this book is made from low- or no-chlorine pulp and is acid free, in conformance with international standards for paper permanency.

Made in the USA
Monee, IL
07 July 2026

56549929R00063